Lecture Notes on
Phase Transformations
in Nuclear Matter

Lecture Notes on
Phase Transformations in Nuclear Matter

Jorge A. López
Univ. of Texas at El Paso, USA

Claudio O. Dorso
Univ. de Buenos Aires, Argentina

World Scientific
Singapore • New Jersey • London • Hong Kong

Published by

World Scientific Publishing Co. Pte. Ltd.

5 Toh Tuck Link, Singapore 596224

USA office: 27 Warren Street, Suite 401-402, Hackensack, NJ 07601

UK office: 57 Shelton Street, Covent Garden, London WC2H 9HE

British Library Cataloguing-in-Publication Data
A catalogue record for this book is available from the British Library.

LECTURES NOTES ON PHASE TRANSFORMATIONS IN NUCLEAR MATTER

ISBN-13 978-981-02-4007-3
ISBN-10 981-02-4007-4

En este trabajo intervinieron:
Mi familia: Rosa Elena, Maribel y Oscar,
   quienes me dieron su tiempo y su amor.
Mis maestros: Roberto Díaz Molina, Roberto Armijo,
   Alan Dean, y Philip Siemens,
   quienes me guiaron y alentaron.
Y la naturaleza,
   que por ser tan lineal,
   nos puso en el camino correcto
   para empezar a entender su complejidad.
Jorge

A mi familia:
  Alicia, Ezequiel, Federico, Sabrina,
  Elsa, José, Nelly y Beto.
A quienes contribuyeron en mi formación:
  E. S. Hernández y J. Randrup.
Claudio

Preface

Nuclear matter is unique in several ways. The fact that, from the light to the heavy, all nuclei have an almost constant density and binding energy shows the strength of the nuclear force as well as its short range. Nucleons in small nuclei feel roughly the same pull from neighboring nucleons as in large nuclei. Thus the famous way of counting nucleons *"One, Two, Three, Four, Infinity"*; after a few nucleons all nuclei are pretty much alike in terms of density and binding energy.

Constant density implies a constant specific volume per nucleon. What keeps the nucleons at that fixed distance from one another? Naively speaking, it has to be a force produced by a strong potential with a narrow and pointed minimum at the right place. And to have stable nuclei, that potential has to have a repulsive core as well as an attractive short tail. Much like a van der Waals gas.

This book deals with some of the exciting possibilities that a van der Waals nuclear liquid has in terms of thermodynamics and phase transformations. Such nuclear systems can be found in nuclei as well as in astrophysical objects. Its behavior in terms of its response to external forces determines, for instance, the rate of oscillations in monopole excitations, the amount of energy stored as heat, and the kinetic energy released through evaporation of light particles during collisions.

The study of phase transformations of nuclear matter was promoted in the decade of the 80's when it began to be experimentally possible to achieve energetic collisions between heavy nuclei with beam energies in the tens of MeVs per nucleon. Starting with an initially idealized picture, the reactions were studied as a perfect laboratory example of a thermodynamic system.

After these initial simple models explained some of the existing inclusive data, the experiments began to provide detailed information that forced a change of the physical picture of the process. It is now believed that heavy ion reactions leading to fragmentation, although achieving a high degree of thermalization, are dominated mainly by finite-size and collision-induced correlations of nucleon density and momentum.

As of now, there is no complete theory to study such complex processes (*e.g.* non-equilibrium critical phenomena in evolving small quantum systems). Our best hope to understand heavy ion reactions, then, is to start with a simple model and gradually add those ingredients required to reach higher degrees of complexity. This work studies the nuclear fragmentation problem starting from a free nucleon gas, adding next a phenomenological nucleon interaction to study possible phase transitions in infinite and finite systems, and to compare to experimental data.

Restating this plan in further detail, chapter 1 introduces a simple free nucleon gas to develop a baseline model to obtain computational methodology, as well as basic medium information and phenomena, such as temperature scales and nucleon evaporation. Adding a phenomenological binding energy, chapter 2 focuses on the resulting nuclear equation of state and its thermodynamic properties. The liquid-gas phase transition is investigated in more detail in chapter 3, where isothermal and isentropic nucleation processes are explored along with spinodal decomposition and other related topics. The fact that the nucleus is a small system is introduced in chapter 4 along with some of the existing techniques for its study. Finally, and since both authors are theorists, the connection to experimental data is timidly presented in chapter 5.

For completeness, the book closes with a brief summary of commonly used computational techniques and some useful programs. [Some of these programs and other interesting ones can be downloaded from the author's web page at www.utep.edu/physics or by anonymous ftp from zapata.utep.edu.

On the technical aspects of the production of this book, the authors kindly acknowledge help from a variety of sources, including, but not limited to Azael Avalos, Pablo Balenzuela, Mario Borunda, Ariel Chernomoretz, Christian Escudero, Martha Navarro, and Rocío Olave. Jorge López also thanks the warm hospitality of the Universidad de Buenos Aires where part of the manuscript was prepared.

Jorge A. López   Claudio O. Dorso

El Paso, Texas, U.S.A.   Buenos Aires, Argentina

March, 2000   March, 2000

Contents

Chapter 1

The Free Nucleon Gas

1.1 Introduction

In its simplest model, a nucleus can be represented as a collection of neutrons and protons in a potential well [Segrè (1982)]. When the number of nucleons is not too small, the volume is restrictive enough to warrant the use of statistical mechanics. In this first approximation, the nucleus is treated as an aggregate of nucleons interacting only through the exclusion principle. This simple model, which will allow us to set the formulae, nomenclature, mass and energy scales to be used throughout the book, it also gives a good representation of the nucleus at low energies, as shown by the Fermi-type energy distributions of nucleons extracted from low energy electron-nucleus scattering [Bohr and Mottelson (1973)].

1.2 The Free Nucleon Gas

For a set of N indistinguishable nucleons in a volume V, at a temperature T, and in a potential well with energy states $\epsilon_1, \ldots, \epsilon_k$, and occupational numbers $n_1, \ldots, n_k, \ldots$, the canonical partition function is given by [McQuarrie (1973)]

$$
*Q(N, V, T) = \sum_{i,j,k,\ldots} e^{-\beta(\epsilon_i + \epsilon_j + \cdots)} = \sum_{\{n_k\}}^{*} e^{-\beta \sum_i \epsilon_i n_i} ,
$$

where $\beta = 1/T$ and the $\sum^*$ indicates that the sum must be performed with the awkward restriction of keeping the total number of particles $N = \sum_k n_k$

fixed. [In nuclear physics it is customary to express the temperature T directly in energy units of MeV, formally this is equivalent to $k_B T$ where T is the temperature in Kelvin.]

To proceed, and since the sum $\sum^*$ is difficult to complete, it is convenient switch to the grand canonical ensemble where V, T, and the chemical potential μ are the independent variables. The grand canonical partition function is given by

$$\Xi(V,T,\mu) = \sum_{N=0}^{\infty} e^{-\beta\mu N} Q(N,V,T) = \sum_{N=0}^{\infty} \sum_{\{n_k\}}^{*} \prod \lambda e^{(-\beta\epsilon_k)^{n_k}} \ ,$$

where $\lambda = e^{\beta\mu}$. Now, since the nucleons are fermions, $n_k = 0,1$ and the partition function reduces to $\Xi_{FD} = \prod_k (1 + \lambda e^{-\beta\epsilon_k})$.

With this, one can now extract many thermodynamic variables of interest. The average number of particles, for instance, is given by

$$N = \sum_k \bar{n}_k = T \left(\frac{\partial \ln \Xi}{\partial \mu} \right)_{V,T} = \sum_k \frac{\lambda e^{-\beta\epsilon_k}}{1 + \lambda e^{-\beta\epsilon_k}} \ , \tag{1.1}$$

i.e. the average occupation number is $\bar{n}_k = \lambda e^{-\beta\epsilon_k} / (1 + \lambda e^{-\beta\epsilon_k})$. Likewise, one can obtain the average total energy, pressure and entropy:

$$E_F = \sum_k \bar{n}_k \epsilon_k = \sum_k \lambda \epsilon_k e^{-\beta\epsilon_k} / (1 + \lambda e^{-\beta\epsilon_k}) \ , \tag{1.2}$$

$$pV = T \ln \Xi(V,T,\mu) = T \sum_k \ln(1 + \lambda e^{-\beta\epsilon_k}) \ , \tag{1.3}$$

$$S = \left[\frac{\partial pV}{\partial T} \right]_{V,\mu} = \sum_k \left[\ln(1 + \lambda e^{-\beta\epsilon_k}) - \frac{(\mu - \epsilon_k)e^{\beta(\mu - \epsilon_k)}}{T(1 + \lambda e^{-\beta\epsilon_k})} \right] . \tag{1.4}$$

In the nuclear case the energies ϵ_k's are almost continuously distributed (*cf.* problem 1.4), and one can go from $\sum_k$ to $\int_\epsilon \omega(\epsilon)d\epsilon$, where $\omega(\epsilon)d\epsilon$ is the number of states with energy between ϵ and $\epsilon + d\epsilon$. Using $\omega(\epsilon)$ from problem 1.1 immediately yields

$$N = 2g\pi (\frac{2m}{h^2})^{3/2} V \int_0^\infty \frac{\lambda \epsilon^{1/2} e^{-\beta\epsilon} d\epsilon}{1 + \lambda e^{-\beta\epsilon}} \ , \tag{1.5}$$

$$E_F = 2g\pi (\frac{2m}{h^2})^{3/2} V \int_0^\infty \frac{\lambda \epsilon^{3/2} e^{-\beta\epsilon} d\epsilon}{1 + \lambda e^{-\beta\epsilon}} \ , \tag{1.6}$$

and

$$p = 2g\pi T(\frac{2m}{h^2})^{3/2}\int_0^\infty \epsilon^{1/2}\ln(1+\lambda e^{-\beta\epsilon})d\epsilon$$
$$= \frac{4}{3}g\pi(\frac{2m}{h^2})^{3/2}\int_0^\infty \frac{\epsilon^{3/2}d\epsilon}{(1+\lambda e^{\beta\epsilon})} = \frac{2}{3V}E_F , \qquad (1.7)$$

where the ln integral in p was reduced using integration by parts. For the nuclear case $g = 4$ to take into account the spin and isospin degrees of freedom. In most cases, the evaluation of expressions such as (1.5)-(1.7) needs the use of numerical integration. In some limits, however, it is possible to approximate the Fermi integrals, this is done for the nuclear case in the following section.

Problem 1.1   The Energy Level Density

(A) *Consider a three dimensional well with energy levels $\epsilon = gh^2(n_x^2 + n_y^2 + n_z^2)/8mV^{2/3}$, where $\{n_i\}$ and g are the quantum numbers and degeneracy, respectively. Using geometrical arguments in n_i-space, show that the number of states available to a particle with energy $\leq \epsilon$ is*

$$\Phi(\epsilon) = \frac{\pi}{6}\left[\frac{8mV^{2/3}\epsilon}{h^2}\right]^{3/2} .$$

(B) *Likewise, show that the number of states in a shell of energy ϵ and thickness $d\epsilon$ is*

$$\omega(\epsilon)d\epsilon = 2g\pi\left(\frac{2m}{h^2}\right)^{3/2}V\epsilon^{1/2}d\epsilon . \qquad (1.8)$$

1.3   Strongly Degenerate Nucleon Gas

For excitations achieved in intermediate-energy reactions, one can argue that the system maintains high level occupancy and is never weakly degenerate (*cf.* problem 1.9). In this limit where the density is high (near saturation density n_0) and or the temperature low ($T \ll \epsilon_F$, *cf.* problem 1.2), most of the states are occupied and $\epsilon \approx \mu$. Expressions (1.5)-(1.7) can be simplified through an expansion in terms of $(\epsilon - \mu)$.

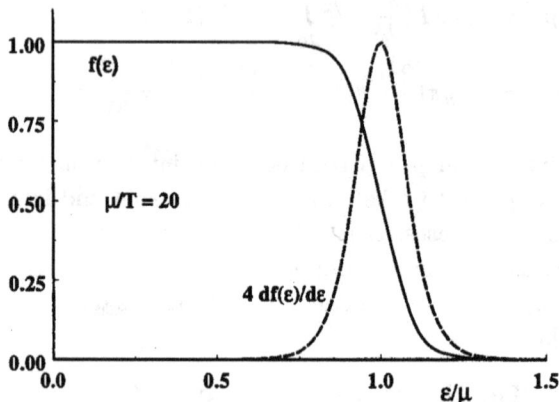

Fig. 1.1   Behavior of $f(\epsilon)$ and $df/d\epsilon$ as a function of ϵ/μ

Expression (1.1) shows that $\bar{n}_k = 1/(1 + e^{\beta(\epsilon_k - \mu)})$, is the probability that state k is occupied. For continuous energy distributions this probability is denoted by $f(\epsilon) = 1/(1 + \lambda e^{\beta\epsilon})$. Figure 1.1 shows the energy dependence of $f(\epsilon)$ as a function of ϵ/μ, clearly $f(\epsilon) \approx 1$ for $\epsilon \le \mu$ and $f(\epsilon) = 0$ for $\epsilon \ge \mu$. That is, all states with energy $\epsilon \le \mu$ are occupied and all other are empty. This helps to obtain simple analytic expressions for the zero temperature case.

Problem 1.2   The T = 0 Case and the Fermi Energy
(A) *Integrate (1.5) to show that in the zero-temperature limit N is given by $N = (4g\pi/3)(2m/h^2)^{3/2}V\mu_0{}^{3/2}$, where μ_0 is the $T = 0$ value of the chemical potential, known as the Fermi energy, ϵ_F. Likewise show that the $T = 0$ value of the total energy of the free nucleon gas is given by $E_0 = 3N\mu_0/5$, and the pressure by $p_0 = 2N\mu_0/(5V)$.*
(B) *Extract the Fermi energy ϵ_F in terms of N, and evaluate it for the nuclear case where $N/V \approx 0.15\ fm^{-3}$ and $m \approx 1000\ MeV/c^2$. Estimate E_0 and p_0 for the nuclear case too.*

For low temperature ($T \ll \epsilon_F$) analytic expressions can be obtained through the use of approximations, here the case of the pressure is worked as an illustration. The integral of expression (1.7) can be evaluated by parts to obtain

$$I = \int_0^\infty \frac{\epsilon^{3/2} d\epsilon}{(1 + \lambda e^{\beta \epsilon})} = \frac{2\beta}{5} \int_0^\infty \frac{\epsilon^{5/2} e^{\beta(\epsilon - \mu)} d\epsilon}{(1 + e^{\beta(\epsilon - \mu)})^2} .$$

This last integral cannot be evaluated analytically, but, as seen in figure 1.1, the factor $df(\epsilon)/d\epsilon = e^{\beta(\epsilon - \mu)}/(1 + e^{\beta(\epsilon - \mu)})^2$ is strongly peaked around $\epsilon \approx \mu$ for low values of T, and one can thus use the expansion $\epsilon^{5/2} = \mu^{5/2} + (\epsilon - \mu)(\partial \epsilon^{5/2}/\partial \epsilon)_{\epsilon = \mu} + (1/2)(\epsilon - \mu)^2 (\partial^2 \epsilon^{5/2}/\partial \epsilon^2)_{\epsilon = \mu} + \cdots$ to obtain

$$
\begin{aligned}
I &= \frac{2\beta \mu^{5/2}}{5} \int_0^\infty \frac{e^{\beta(\epsilon - \mu)} d\epsilon}{(1 + e^{\beta(\epsilon - \mu)})^2} + \beta \mu^{3/2} \int_0^\infty \frac{(\epsilon - \mu) e^{\beta(\epsilon - \mu)} d\epsilon}{(1 + e^{\beta(\epsilon - \mu)})^2} \\
&+ \frac{3\beta \mu^{1/2}}{4} \int_0^\infty \frac{(\epsilon - \mu)^2 e^{\beta(\epsilon - \mu)} d\epsilon}{(1 + e^{\beta(\epsilon - \mu)})^2} + \cdots .
\end{aligned}
$$

The first integral equals simply $2\beta \mu^{5/2}/5$, and to approximate the rest of the integrals one extends the lower limit of integration to $-\infty$. All those integrals with odd powers of $(\epsilon - \mu)$ have odd integrands and are thus equal to zero. The remaining integrals are given by $I_i = \int_{-\infty}^\infty t^i e^t dt/(1 + e^t)^2 = 2i(i-1)!(1-2^{1-i})\zeta(i)$, for i integer and even, and where $\zeta(i)$ is the Riemman zeta function. The degenerate limit of the pressure is then given by

$$p = \frac{8g\pi}{15} \left(\frac{2m}{h^2} \right)^{3/2} [\mu^{5/2} + 5\pi^2 T^2 \mu^{1/2} + \cdots] . \tag{1.9}$$

A similar derivation for N gives

$$N = \frac{8g\pi}{3} \left(\frac{2m}{h^2} \right)^{3/2} V[\mu^{3/2} + \frac{\pi^2}{8} T^2 \mu^{-1/2} + \cdots] . \tag{1.10}$$

The corresponding expression for E_F can be obtained from equation (1.7), and (1.9). The entropy can be obtained from these previous expressions and the fundamental Gibbs-Duhem equation is $S = E_F/T - N\mu/T + pV/T$.

Problem 1.3　Strongly Degenerate Expression for N
Derive expression (1.10) starting from (1.5). Use it to sketch the behavior of the number density $n = N/V$ for the range 1 MeV $< T <$ 10 MeV and 20 MeV $< \mu <$ 40 MeV.

So far μ and T have been used as the independent variables, but sometimes it is more convenient to use the density and temperature instead. This can be achieved by inverting expression (1.10) to obtain the chemical potential as a function of T and ϵ_F (see *eg.* [McQuarrie (1973)]):

$$\mu = \epsilon_F \left[1 - \frac{\pi^2}{12} \left(\frac{T}{\epsilon_F} \right)^2 - \frac{\pi^4}{180} \left(\frac{T}{\epsilon_F} \right)^4 + \cdots \right] . \qquad (1.11)$$

This can be used to obtain

$$E_F = \frac{3}{5} N \epsilon_F \left[1 + \frac{5\pi^2}{12} \left(\frac{T}{\epsilon_F} \right)^2 - \frac{\pi^4}{16} \left(\frac{T}{\epsilon_F} \right)^4 \cdots \right] . \qquad (1.12)$$

The thermal energy $\epsilon_T(n, T) = (E_F(n, T) - E_F(n, 0))/N$ as obtained with approximation (1.12) is compared in figure 1.2 to a numerical evaluation and a quadratic fit that will be introduced in the next sections.

The heat capacity can also be calculated as

$$C_V = \left(\frac{\partial E_F}{\partial T} \right)_V = \frac{\pi^2 N k^2 T}{2\epsilon_F} + \cdots . \qquad (1.13)$$

These expressions can be evaluated using the result of problem 1.2, namely, $\epsilon_F = (\hbar^2/2m)(3N/8g\pi V)^{2/3}$.

1.3.1   *Energy Level Density*

In the present free nucleon gas approximation, the only interaction between nucleons is that of the exclusion principle. This can be highly restrictive for the particle dynamics at low temperatures. To quantify the number of available energy states, the previously derived strongly degenerate expressions will be used to obtain a crude estimate of the density of energy states.

From the statistical and thermodynamic definitions of entropy,

$$S(E_F(T)) = \log \left[\frac{\omega(E_F(T))}{\omega(E_F(0))} \right] , \quad S(E_F(T)) = \int_0^T dE_F(T)/T ,$$

and using equation (1.12), one obtains $S = N\pi^2 T/2\epsilon_F$. Now, defining the level density parameter as $a = N\pi^2/4\epsilon_F$ and since $E_T(T) = N\pi^2 T^2/4\epsilon_F = aT^2$, the entropy can be written as $S = 2\sqrt{aE_T}$. With this the energy level density is

$$\omega(E_F(T)) = \omega(E_F(0))e^{2\sqrt{aE_T(T)}} . \qquad (1.14)$$

It is worth noticing that in this derivation the entropy does not contain contributions from volume and particle number changes, *i.e.* it corresponds to the case of fixed volume and N.

The ratio $\omega(E_F(T))/\omega(E_F(0))$ can be taken as a measure of the opening of the phase-space as a function of E_T. For instance, as shown in problem 1.4, the number of available states for 100 nucleons increases by a factor of a million at energies as E_T goes from zero to 4 MeV. The restrictions imposed by Pauli's principle should be of no concern in the nuclear case.

Problem 1.4   Nuclear Level Density
(A) *Fill in the details in the derivation of expression (1.14).*
(B) *Evaluate the level density parameter a for the nuclear case and show that $a \sim N/8$ at n_0.*
(C) *Sketch the ratio $\omega(E_F(T))/\omega(E_F(0))$ for $E_F = 7\,MeV$ and the range $10 < N < 240$. [You can compare your gross estimate to the experimental data of figure 11.29 of [Segrè (1982)] where $\omega(E_F(0))$ and a have been adjusted to fit the data.]*

1.4   Numerical Evaluations

For the general case, expressions (1.5)-(1.7) must be evaluated numerically. Letting $g = 4$ and $\eta = \mu/T$, expressions (1.5) and (1.7) can be written as

$$n = \frac{N}{V} = \frac{1}{\pi^2}\left(\frac{2mc^2}{(\hbar c)^2}\right)^{3/2} T^{3/2}\int_0^\infty \frac{t^{1/2}dt}{1+e^{t-\eta}}$$

$$p = \frac{2}{3\pi^2}\left(\frac{2mc^2}{(\hbar c)^2}\right)^{3/2} T^{5/2}\int_0^\infty \frac{t^{3/2}dt}{1+e^{t-\eta}}.$$

The Fermi integral $\int_0^\infty t^i dt/(1+e^{t-\eta})$ and can be evaluated using any standard package, such as the *FERDR* subroutine $C323$ of the *CERN* library (conveniently reproduced in appendix B.2). Furthermore, to eliminate the dependence on the number of particles, the energy per particle can be evaluated from

$$\varepsilon_F = E_F/N = \frac{3p}{2n} = T\frac{\int_0^\infty t^{3/2}dt/(1+e^{t-\eta})}{\int_0^\infty t^{1/2}dt/(1+e^{t-\eta})}.$$

The numerical evaluation of the thermal energy of the free nucleon gas, $\varepsilon_T(n,T) = (E_F(n,T) - E_F(n,0))/N$, is compared in figure 1.2 to the

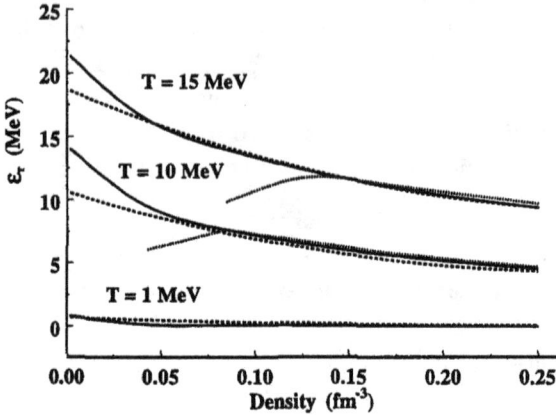

Fig. 1.2 Behavior of a numerical integration (solid), quadratic fit (dashed) and the strongly degenerate expansion (dotted) of $\varepsilon_T(n,T) = (E_F(n,T) - E_F(n,0))/N$ as a function of the density n and for three different temperatures.

strongly degenerate expansion of section 1.3, and to a fit to be introduced in the next section.

Problem 1.5   Numerical Evaluation of n, p, ε and s
Use code Fermigas.for *from appendix B.1 and subroutine* Ferdr.for *from appendix B.2 to evaluate numerically n, p, ε and the entropy per particle, $s = S/N$, for the ranges $1\,MeV < T < 10\,MeV$ and $20\,MeV < \mu < 40\,MeV$. Compare your number density results to those of problem 1.3.*

1.5   Useful Fits

Although a numerical integration of the thermodynamic expressions of the free nucleon gas is more generically applicable, it is possible to obtain fits to these expressions which are valid within a specific range of the independent variables. For instance, the temperature part of the energy per particle, namely $\varepsilon_T(n,T) = (E_F(n,T) - E_F(n,0))/N$ can be fitted by a doubly

Table 1.1 Coefficients ε_{ij}^T and ε_{ij}^S

Coefficient	Value	Coefficient	Value
ε_{01}^T	0.693	ε_{02}^T	0.037
ε_{11}^T	−5.420	ε_{12}^T	0.082
ε_{21}^T	11.447	ε_{22}^T	−0.312
ε_{01}^S	−0.0984	ε_{02}^S	0.845
ε_{11}^S	−2.481	ε_{12}^S	20.607
ε_{21}^S	−1.697	ε_{22}^S	17.151

quadratic polynomial of the form

$$\varepsilon_T(n, T) = \sum_{i=0}^{2} \varepsilon_{Ti}(T) n^i \, ,$$

with the temperature-dependent coefficients given by $\varepsilon_{Ti}(T) = \sum_{j=1}^{2} \varepsilon_{ij}^T T^j$.

A similar fit can be obtained if ε is needed in terms of the entropy: $\varepsilon_S(n, S) = \sum_{i=0}^{2} \varepsilon_{Si}(S) n^i$ with the entropy-dependent coefficients given by $\varepsilon_{Si}(S) = \sum_{j=1}^{2} \varepsilon_{ij}^S S^j$. Table 1.1 lists the values of the coefficients $\{\varepsilon_{ij}^T\}$ and $\{\varepsilon_{ij}^S\}$ that best reproduce ε_T and ε_S, respectively for the ranges of interest in the nuclear case: $0 < T < 20$ *MeV*, $0 < S < 3$, and $0 < n < 0.2$ *fm*$^{-3}$. [Notice that in the present units S is unitless, *i.e.* it is measured in units of Boltzmann's constant.] Figure 1.2 compares the temperature dependent fit of $\varepsilon_T(n, T)$ to the degenerate expression and the numerical evaluation presented in sections 1.3 and 1.4, respectively.

1.5.1 *Caloric Curve*

The fits obtained for the thermal energy, ε_T, can help us study the so-called *caloric* curve. This the existing relationship between the temperature of the free nucleon gas and its thermal energy.

Since the thermal energy, $\varepsilon_T(n, T)$, depends explicitly on T through a simple doubly quadratic polynomial, the caloric curve can be obtained by inverting this expression for T. This procedure yields

$$T(n, \varepsilon_T) = -T_B(n) + \frac{\sqrt{T_B^2(n) + 4\, T_A(n) \varepsilon_T}}{2\, T_A(n)} \, ,$$

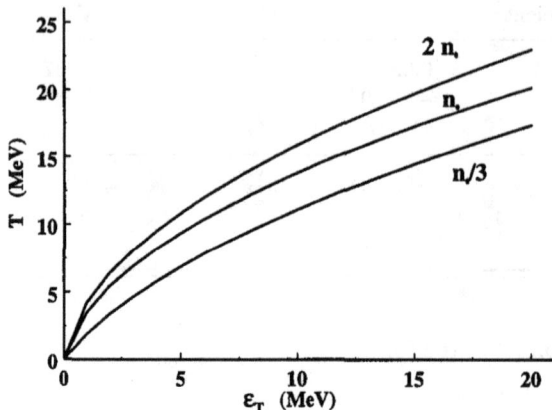

Fig. 1.3   Caloric curve for the free nucleon gas at constant densities

where $T_A(n) = \sum_{i=0}^{2} \varepsilon_{i2}^T n^i$ and $T_B(n) = \sum_{i=0}^{2} \varepsilon_{i1}^T n^i$. Figure 1.3 shows the $T - \varepsilon_T$ relationship at three values of the density going from one third to twice saturation density.

For real gases, the caloric curve (at constant pressure) has the peculiarity of being piecewise continuous at phase changes. As the present model does not contain nucleon-nucleon interactions, its caloric curve does not present these features. For a comparison to a caloric curve extracted from experimental data see figure 4.10 in section 4.5 or figure 5.4 in section 5.5.

1.6   Nucleon Evaporation

Although, due to the lack of particle-particle interactions, the free nucleon gas does not present phases, it can *"evaporate"* particles. Evaporation in this context refers to the separation of particles from the main body of the gas. In this section, as an application of the free nucleon gas model, two alternative nucleon radiation techniques will be studied: thermionic emission and Weisskopf evaporation. For a comparison of these methods, see [Friedman and Lynch (1983)].

1.6.1 *Thermionic Emission*

Consider a gas of free nucleons contained in a square well potential of height W, with the property $W > \epsilon_F$. The momentum distribution is given by $f(\epsilon)$ (*cf.* section 1.3) expressed in terms of the momentum $\vec{p}$: $f(\vec{p}) = 1/(1 + e^{(\vec{p}^2/2m-\mu)/T})$. If the nucleons are contained in a region of space separated from the vacuum by a surface, those particles in the high-momentum end of the distribution will be able to escape into the vacuum.

Placing the surface perpendicularly to the z-direction, particles with $p_z^2/2m > W$ will escape in a unit time provided they are at a distance $dz = v_z$ from the surface, where v_z is the velocity perpendicular to the surface. The number of particles escaping per unit time per unit surface is then given by

$$R = \int_{p_z=\sqrt{2mW}}^{\infty} \int_{p_{xy}} d^3p \, n_p \, v_z \, f(\vec{p}) \, ,$$

where p_{xy} is the component of the momentum transverse to the z direction, $v_z = p_z/m$, and n_p is the number of quantum states per unit volume, *i.e.* $n_p = (gV/h^3)/V = g/h^3$. Using $d^3p = dp_z dp_{xy} d\theta p_{xy}$

$$\begin{aligned}
R &= \int_{p_z=\sqrt{2mW}}^{\infty} \int_0^{\infty} \int_0^{2\pi} \frac{g}{mh^3} \frac{d\theta \, p_{xy} dp_{xy} \, p_z dp_z}{(1 + e^{((\vec{p_z}^2 + \vec{p_{xy}}^2)/2m - \mu)/T})} \, , \\
&= \frac{2g\pi mT}{h^3} \int_{\epsilon_z=W}^{\infty} d\epsilon_z \ln\left[1 + e^{(\mu - \epsilon_z)/T}\right] \, ,
\end{aligned}$$

where a change of variable to $\epsilon_z = \vec{p_z}^2/2m$ has been introduced. Now, for low temperatures ($T \ll W$), and since $W > \mu$ the exponential term will be small and thus

$$R \approx \frac{2g\pi mT}{h^3} \int_W^{\infty} d\epsilon_z e^{(\mu-\epsilon_z)/T} = \frac{2g\pi mT^2}{h^3} e^{(\mu-W)/T} \, . \tag{1.15}$$

Since μ is a function of the temperature and density, R depends effectively on these variables too.

Problem 1.6 Thermionic Emission

Use expression (1.15) to calculate the number of nucleons that radiate away from a $T = 3\,MeV$, $N = 100$ nucleus in a time interval of 10^{-21} sec. Take $\mu - W$ equal to the separation energy $8\,MeV$, and estimate the nuclear surface taking the nucleus as a sphere of radius $r = r_0 N^{1/3}$ with $r_0 = 1.2\,fm$.

The simple free nucleon gas used here does not distinguish between neutrons and protons. In a more realistic model (see *e.g.* [López and Randrup (1994)]), escaping protons would feel a Coulomb repulsion of the form $W_{protons} = e^2(Z-1)/r$ (where Ze and r are the nuclear charge and radius), whereas for the neutrons, $W_{neutrons} = 0$. It is worth mentioning that the product $R \times Surface \times Time$ is a dN/dt, and as such it should be integrated when calculating the emission over long periods of time as N and T might change .

1.6.2 *Weisskopf Evaporation*

An alternative, but equivalent way of calculating the emission of nucleons from an excited free nucleon gas comes from the statistical theory of nuclei [Weisskopf (1937)]. According to this theory, the rate of decay (*i.e.* decays per unit time) of a nucleus "C" into a nucleon "b" and a daughter nucleus "B" is given by

$$\frac{dN^2}{dKdt} = \frac{gm}{\pi^2\hbar^3}\,\sigma^c_{b+B}K\frac{\omega_B(E_B)}{\omega_c(E_c)} \ , \tag{1.16}$$

where m, K, and g refer to the mass, kinetic energy and degeneracy of the emitted nucleon, $\omega_B(E_B)$ and $\omega_c(E_c)$ are the level densities of the daughter and mother nuclei at *thermal* energies E_B and E_c, respectively, and σ^c_{b+B} is the $b + B \rightarrow c$ reaction cross section.

To integrate over all values of K, a level density appropriate for the case when N is not held constant will be used. Keeping the volume constant, and proceeding as in section 1.3.1, the entropy and level density are given by

$$S = \int_0^T dE/T - \int_0^N (\mu/T)dN = 2\sqrt{aE_T} - (\mu/T)N \ \ ,$$
$$\omega(E) = \omega(0)e^{2\sqrt{aE_T}-(\mu/T)N} \ .$$

Now since $E_c = E_B + K + W$, and $N_c = N_B + 1$, the ratio in (1.16) can be expressed as

$$\frac{\omega_B(E_B)}{\omega_c(E_c)} = e^{2\sqrt{aE_c}\left(\sqrt{1-(K+W)/E_c}-1\right)+\mu/T} \approx e^{-(K+W)/T+\mu/T} \ ,$$

where $\sqrt{1 - (K + W)/E_c}$ has been expanded to first order, and $\sqrt{a/E_c}$ has been replaced by T as in section 1.3.1. The nucleon emission per unit time is thus

$$\frac{dN}{dt} = \frac{g \, m \, \sigma^c_{b+B}}{\pi^2 \hbar^3} \int_0^{E_c+W} K e^{-(K+W)/T + \mu/T} dK \approx \frac{g \, m \, \sigma^c_{b+B}}{\pi^2 \hbar^3} T^2 e^{(\mu-W)/T} \, ,$$

where, taking advantage of the fast decreasing exponential, the upper limit of integration has been replaced by ∞. Taking the emitting surface as $4\sigma^c_{b+B}$, the number of particles escaping per unit time per unit surface is given by

$$R = \frac{2 \, g \, \pi \, m}{h^3} T^2 e^{(\mu-W)/T} \, , \tag{1.17}$$

in agreement with the thermionic emission expression (1.15). Expression (1.17) was derived keeping the volume fixed, reference [Friedman and Lynch (1983)] extends the present approach to the case of constant density.

1.7 Some Conclusions

The free nucleon gas is the simplest model of nuclear matter. In this case, statistical mechanics can be applied due to the relatively large number of nucleons for the restrictively small volume. Using only the exclusion principle, the model represents the nucleus as a collection of neutrons and protons in a potential well with a density of energy levels not too different from the real one. Although computationally demanding, the model can be approximated by the use of numerical expansions, integrations and fits. The nuclear scales of interest that result from this model are: $0 < T < 20 \, MeV$, $0 < S < 3$, $0 < n < 0.2 \, fm^{-3}$, $20 \, MeV < \mu < 40 \, MeV$, and $0 < \varepsilon_T < 40 \, MeV$ with a value of N in the hundreds of particles and $n_0 = 0.15 \, fm^{-3}$. In terms of dynamical phenomena, this simple model predicts the possibility of nucleon evaporation.

1.8 Additional Problems

Problem 1.7   Obtaining N, E_F, and p
Obtain N, E_F and p (expressions (1.5)-(1.7)) starting from equations (1.1)-(1.3) and using expression (1.8).

Problem 1.8   Entropy
Show that $S = E_F/T - N\mu/T + pV/T$ starting from expressions (1.4) and (1.8).

Problem 1.9   Weak Degeneracy
Weak degeneracy sets in when the de Broglie wavelength, Λ, of the nucleons is roughly of the size of the nucleon's volume, i.e. $\Lambda^3 \approx V/N$, where $\Lambda = \sqrt{h^2/(2\pi m T)}$. For weak degeneracy, the ratio $N\Lambda^3/V$ would have to be approximately close to unity. Sketch the behavior of $N\Lambda^3/V$ for the nuclear case using $N/V = 0.15\ fm^{-3}$, $m = 938\ MeV/c^2$, and T ranging from 0 to 20 MeV's. What is a "safe" temperature limit for weak degeneracy?

Problem 1.10   Strongly Degenerate Expression for μ and E
Derive expressions (1.11)-(1.13). Hint: see section 10.2 in [McQuarrie (1973)].

Problem 1.11   Helmholtz Free Energy
Show that the strongly degenerate expression for the Helmholtz free energy is given by $A = (3N\epsilon_F/5)[1 - (5\pi^2/12)(T/\epsilon_F)^2 + (\pi^4/48)(T/\epsilon_F)^4 \cdots]$.

Problem 1.12   Strongly Degenerate Expression for S
Show that the strongly degenerate expression for the entropy is given by $S = (N\epsilon_F/T)[(\pi^2/2)(T/\epsilon_F)^2 - (\pi^4/20)(T/\epsilon_F)^4 \cdots]$.

Chapter 2

A Simple Nuclear Equation of State

In the free nucleon gas the particles interact with one another only through the exclusion principle embedded in the Fermi-Dirac statistics. In real nuclei, however, the strong and Coulomb forces also play an important role in the stability of the nucleus. To build a working model of the nucleus, the effects of the interactions must be added to the free nucleon gas. As there is not yet a first-principles way of solving the many body problem with strong interactions, it is necessary to resort to a phenomenological approach.

Most models simply add an interaction energy to the free nucleon gas. In order to represent a nucleus, the energy per nucleon at $T = 0$ should have a value of $\varepsilon(n_0) = -8$ MeV at saturation density n_0. Also, since a nucleus is stable at this density, the pressure should vanish, and as observed in experiments of nuclear oscillations [Youngblood et $al.$ (1984)], the compressibility should be of the order $100\,MeV < K(n_0) < 300\,MeV$. Several parametrizations of the interaction energy are able to reproduce these values ($eg.$ [Danielewicz (1979)], [Bertsch and Siemens (1984); López and Siemens (1984)], [Kapusta (1984)], and others [Goodman et $al.$ (1984)]), for simplicity the equation of state proposed by Kapusta will be used with the doubly quadratic fit of section 1.5 to mimic the temperature dependence of the free nucleon gas component.

2.1 The Nuclear Equation of State

The equation of state (EOS) of a medium formally relates the pressure with two independent variables, commonly taken as the density and the temperature. This expression for $p(n, T)$, can be obtained from the energy of the system through $p(n,T) = n^2(\partial \varepsilon(n,T)/\partial n)_S$. Adding an interaction energy $\varepsilon_{int}(n)$ to the free nucleon gas, the energy per nucleon (measured with respect to the $T = 0$ value) is thus given by

$$\varepsilon(n,T) = \varepsilon_{int}(n) + [\varepsilon_F(n,T) - \varepsilon_F(n,0)] = \varepsilon_0(n) + \varepsilon_T(n,T) .$$

Notice that in this simple model, the interaction energy per particle is assumed to be temperature independent and denoted by $\varepsilon_{int}(n) = \varepsilon_0(n)$. The thermal part of the energy per nucleon, ε_T, is given in terms of the free nucleon gas energy, $\varepsilon_F = E_F/N$, (*cf.* Eq. (1.6)) and its zero-temperature value, $\varepsilon_F(n,0) = 3\mu_0(n)/5$ (*cf.* problem 1.2).

The parametrization introduced by [Kapusta (1984)] sets the ground state energy as $\varepsilon_0(n) = \sum_{i=2}^{5} a_i(n/n_0)^{i/3}$, with the parameters $\{a_i\}$ adjusted to yield $p = 0$, $\varepsilon_0 = -8 \ MeV$, $K = 210 \ MeV$ and the correct zero-temperature value of the Fermi gas energy, all at saturation density. [The isothermal compressibility is obtained from $K = n(\partial p/\partial n)_T$.] The corresponding equation of state is thus

$$p(n,T) = \frac{n_0}{3} \sum_{i=2}^{5} i a_i \left(\frac{n}{n_0}\right)^{i/3+1} + \varepsilon_T(n,T) , \qquad (2.1)$$

with $a_2 = 21.1$, $a_3 = -38.3$, $a_4 = -26.7$, and $a_5 = 35.9$ all in units of MeV. To further simplify this expression, one can use the fit presented in the previous section, namely $\varepsilon_T(n,T) = \sum_{i=0}^{2} \varepsilon_{Ti}(T)n^i$ with the temperature-dependent coefficients of ε_{Ti} listed in table 1.1. Figure 2.1 shows the isothermal pressure curves obtained with equation (2.1) as a function of the number density.

As seen in figure 2.1, the pressure isotherms have a behavior characteristic of classical media. Normally, a reduction of the volume occupied by a gas, *i.e.* an increase of the density, increases the pressure since $\partial p/\partial n > 0$. Figure 2.1, however, shows the existence of a region where $\partial p/\partial n < 0$. In this region the compressibility is negative, and the system responds to an increase of the density with a reduction of the pressure and a further increase of the density. This is not a stable region, as those density fluctuations that normally occur in any continuous media will break the system into pieces of low density (gaseous particles) and high density (liquid particles). These regions are known as the isothermal (if $(\partial p/\partial n)_T < 0$) and adiabatic (for $(\partial p/\partial n)_S < 0$) *spinodal* regions.

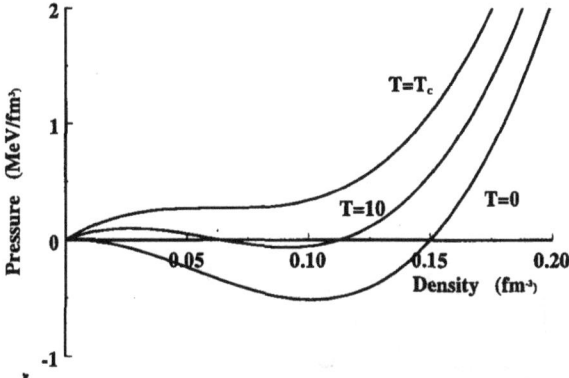

Fig. 2.1 Isothermal pressure curves as a function of the density for three temperatures, with $T_c = 14.542 \, MeV$.

But this is not the whole story, even before a nucleus enters the spinodals, there is a region where, although $\partial p/\partial n > 0$, no extra energy is needed remove a nucleon from a nucleus, or to bind one more to a free nucleon. Inside this region, known as the *coexistence* region, uniform nuclear matter abandons the pressure isotherms and breaks into phases adopting a pressure determined by the temperature and the average density of the system. This section will study the coexistence, isothermal and spinodal regions.

Problem 2.1   The Coefficient a_2
In the strongly degenerate case, the zero-temperature part of the Fermi gas energy goes like $E_F(n, T = 0) \sim n^{2/3}$. Calculate $E_F(n, 0)$ and compare the coefficient of $n^{2/3}$ to the EOS parameter a_2.

2.2   Coexistence Curve

Inside the coexistence region the liquid and gaseous phases can be in equilibrium with one another. When this happens $T_{liquid} = T_{gas}$, $p_{liquid} = p_{gas}$ and $\mu_{liquid} = \mu_{gas}$. Unfortunately these conditions show no special geomet-

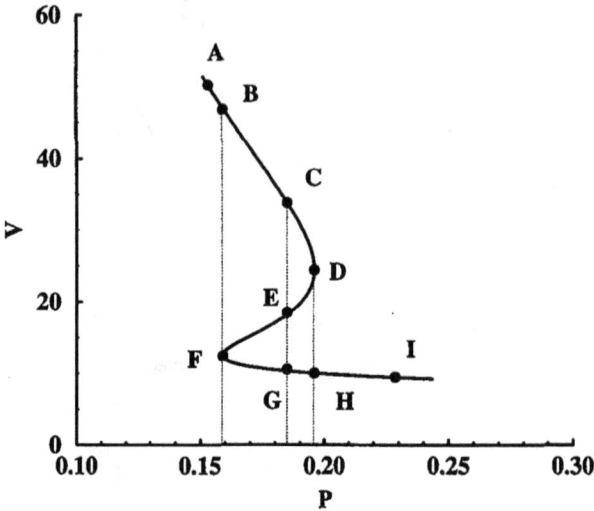

Fig. 2.2   Pressure Isotherm for $T = 13$ MeV plotted sideways versus the volume and showing the unstable region between points D and F.

rical features in the pressure curves and the boundary of the coexistence region must be determined by a special technique known as the *Maxwell construction*. Figure 2.2 shows the pressure isotherm for $T = 13$ MeV plotted sideways against the volume. The unstable region goes from point D to point F. To determine the boundary of the coexistence region it is necessary to look at the energy needed to add or subtract a nucleon from either phase, *i.e.* the Gibbs free energy, $g = \mu = \epsilon - Ts + pV$.

Infinitesimal changes of the Gibbs free energy per particle are given by $dg = -sdT + Vdp$. In an isotherm, where $dT = 0$, the change in the Gibbs free energy per particle is the area below the curve $V(p)dp$. To determine the boundary of the coexistence region, this area must be equal to zero.

Taking figure 2.2 as a reference, the area of interest is the one below the curve. Due to the change in the sign of dp along the loop, $(g_2 - g_1)$ is not a monotonous function of the limits. Then the corresponding expression for our *EOS*, in terms of the temperature and the volume, reads:

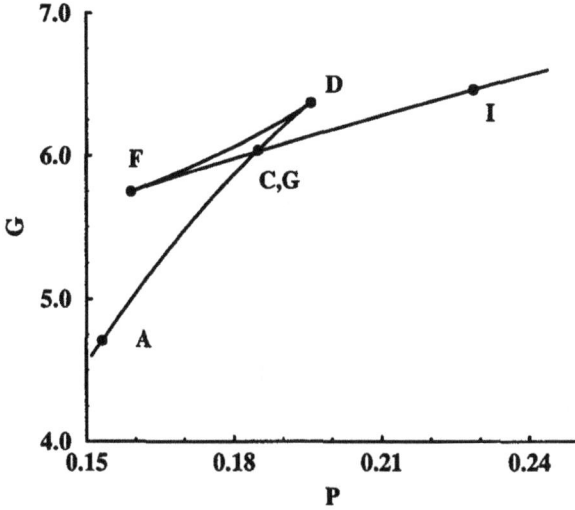

Fig. 2.3 *G* as a function of *p* for our *EOS* at $T = 13\,MeV$, *i.e.* just below T_c

$$g_2 - g_1 = \int_{p_1}^{p_2} V(p)\,dp =$$

$$= \left[\left(\frac{5a_2}{3n_0^{2/3}} + \frac{8a_5}{3n_0^{5/3}V} \right) \frac{1}{V^{2/3}} + \right.$$

$$- \frac{2}{3} \left(\varepsilon_{01}^T + \varepsilon_{02}^T\, T \right)\, T\, \log(V) + \frac{7a_4}{3n_0^{4/3}} \frac{1}{V^{4/3}}$$

$$\left. + \frac{6a_3/n_0 + 4\varepsilon_{11}^T\, T + 4\varepsilon_{12}^T\, T^2}{3V} + \frac{\varepsilon_{21}^T\, T + \varepsilon_{22}^T\, T^2}{V^2} \right]_1^2$$

where the ε_{ij}^T are given in table 1.1.

Figure 2.3 displays the behavior of g as a function of P for our *EOS* at a temperature of 13 MeV, i.e. below T_c. For a system to be in equilibrium g should be a minimum. So the allowed values of g are the ones lying on the curve $ACGI$. The points C and G are coincident, and can be determined

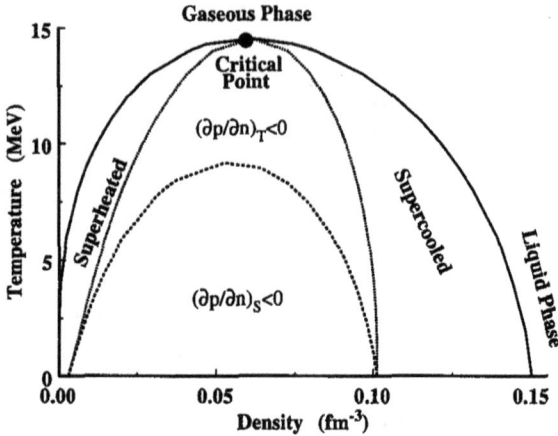

Fig. 2.4   Phase diagram of nuclear matter

by the condition $0 = \int_{p_c}^{p_g} V(p)dp$, which implies that

$$\int_{p_c}^{p_d} V(p)dp - \int_{pe}^{p_d} V(p)dp = \int_{p_f}^{p_e} V(p)dp - \int_{p_f}^{p_g} V(p)dp \ .$$

This states that the areas determined by the curve of constant pressure p and the isotherm under study, to the left and the right of the intersect in the unstable region, must be equal. In this way the set of solutions of this "equal areas" condition determine the so called coexistence curve.

It is clear that the endpoints of the Maxwell construction can also be calculated from the conditions $(p_c = p_g)_{T,p}$ and $(\mu_c = \mu_g)_{T,p}$ since, in equilibrium, the pressures and the chemical potential of both phases are to be equal. The portion of the isotherms from the boundary of the coexistence curve to the limits of the unstable spinodal region are physically accessible, and are called the *superheated vapor* and *supercooled liquid* (*cf.* figure 2.4).

The Mathematica notebook Coexistence.nb listed in appendix B.3, is a pedestrian but illustrative (and fun) way to determine the boundary of the coexistence region.

2.3 Spinodal Regions

The dotted line of figure 2.1 shows the boundary of the unstable region, this is known as the isothermal spinodal line. This curve was calculated solving $(\partial p(n, T)/\partial n)_T = 0$ for n for a number of temperatures. The right branch of the curve shows the points where a further increase of the liquid density results in no change of pressure. The left branch of the curve shows the corresponding points for the gaseous phase.

The left and right branches of the isothermal spinodal meet at the critical point (n_c, T_c). For the particular parametrization of equation (2.1), the critical point occurs at $n_c = 0.061 \ fm^{-3}$ and $T_c = 14.542 \ MeV$.

For adiabatic processes, *i.e.* those conserving the total entropy of the system, the nuclear medium can be described by the entropy counterpart of equation (2.1), $p(n, S) = n_0/3 \sum_{i=2}^{5} i a_i (n/n_0)^{i/3+1} + \varepsilon_S(n, S)$, with $\varepsilon_S(n, S)$ given as in section 1.5. Again, the isentropic pressure $p(n, S)$ has a zone of negative compressibility (now the isentropic compressibility) known as the mechanically unstable region. This boundary, known as the *adiabatic spinodal* line, is outlined in figure 2.4 with a dashed curve. For this particular parametrization the end-point of the spinodal is located at $n = 0.034 \ fm^{-3}$ corresponding to an entropy of $S = 2.145$.

Problem 2.2   Boundary of the Unstable Region
Use the equation of state (2.1) to determine the critical point and several points of the isothermal spinodal up to four significant digits. Hint: the condition $(\partial p(n, T)/\partial n)_T = 0$ yields the requirement:

$$0 = 5a_2 n_0 \ (n/n_0)^{2/3} + 9a_3 n + 14a_4 n \ (n/n_0)^{1/3} + 20a_5 n \ (n/n_0)^{2/3} + 3\varepsilon_{T0} n_0 + 6\varepsilon_{T1} n_0 n + 9\varepsilon_{T2} n_0 n^2.$$

Problem 2.3   The Mechanically Unstable Region
Repeat problem 2.2 to determine the end-point and several points of the adiabatic spinodal. Hint: the condition $(\partial p(n, S)/\partial n)_S = 0$ yields the same requirement as in problem 2.2 with the set $\{\varepsilon_S\}$ replacing $\{\varepsilon_T\}$.

Problem 2.4   The Entropy at T_c
Using the T-dependent equation of state (2.1) and its entropy counterpart, determine the entropy of the critical point. Hint: solve $p(n_c, T_c) = p(n_c, S_c)$ for S_c. What is the temperature of the end-point of the spinodal?

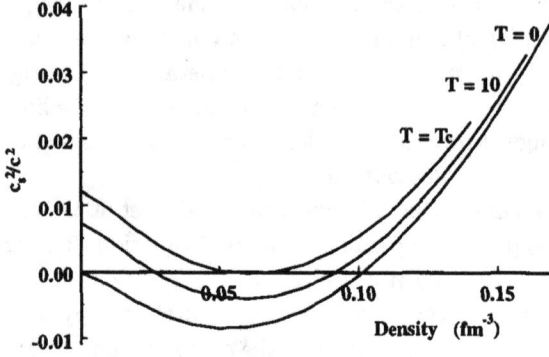

Fig. 2.5   Isothermal speed of sound squared normalized by c^2 as a function of n for $T = 0$, 10, and $T_c = 14.542$ MeV.

2.4   Speed of Sound

The behavior of sound is deeply connected to the spinodal regions. Due to the ill-behaved speed of sound inside the unstable zone, it is better to study the square of the speed of sound, $c_s^2 = (\partial p/\partial n)_T/m$, which for our equation of state is

$$
c_s^2 = \frac{2}{9mn_0} \left[9a_3 n + 14a_4 n \left(\frac{n}{n_0}\right)^{1/3} + 5a_2 n_0 \left(\frac{n}{n_0}\right)^{2/3} + 20a_5 n \left(\frac{n}{n_0}\right)^{2/3} + 3n_0 \varepsilon_{T0} + 6n_0 n \varepsilon_{T1} + 9n_0 n^2 \varepsilon_{T2} \right] \tag{2.2}
$$

where the ε_{Ti} are listed in table 1.1. A couple of observations are in order, first, if m is given in MeV/c^2, c_s^2 will be in units of c^2, the speed of light squared. Second, in deriving expression (2.2), T has been held constant to yield the isothermal speed of sound, the adiabatic speed is given by a similar expression with T replaced by S, and the functions ε_{Ti} by ε_{Si}.

The peculiar behavior of c_s^2 is shown in figure 2.5. For a number of densities and temperatures $c_s^2 < 0$, the region where this happens corresponds to the isothermal spinodal, as discussed in section 2.3.

2.5   More Conclusions

Here the free nucleon gas was dressed up with nucleon-nucleon interactions to build a more realistic model of nuclear matter. The Kapusta parametrization of the interaction energy was coupled with the doubly quadratic fit of section 1.5 to produce an interacting nucleon gas model. Cold nuclear matter at saturation density ended with a binding energy per particle of $\varepsilon(n_0) = -8\ MeV$, zero pressure, and a compressibility of $K(n_0) < 210\ MeV$.

The resulting model allows an easy calculation of isothermal and isentropic pressure curves as a function of the number density. These calculations show the existence of liquid and gaseous phases, as well as a *coexistence* region and unstable isothermal and adiabatic *spinodal* regions. For the equation of state used, the critical point occurred at $n_c = 0.061\ fm^{-3}$ and $T_c = 14.542\ MeV$, and the end-point of the adiabatic spinodal at $n = 0.034\ fm^{-3}$ and $S = 2.145$. The model also allowed an easy parametric calculation of the speed of sound which again showed the unstable regions.

This model will now be used in a more in-depth study of the dynamics of phase changes.

2.6   Additional Problems

Problem 2.5   Saturation Density
Using expression (2.1) verify that nuclear matter is stable at saturation density n_0, i.e. verify that $p = 0$ in the ground state at n_0.

Problem 2.6   Nuclear Compressibility
Derive an expression for the isothermal compressibility K starting from expression (2.1), and verify that $K = 210\ MeV$ at the ground state at saturation density.

Chapter 3

The Road Toward Mixed Phases

Nuclear matter can go from a uniform phase to a mixed phase if it enters the coexistence region (*cf.* figure 2.4). The passage from a continuous phase to a liquid-gas mixture can happen in several ways depending on the initial density and temperature (or entropy) of the nucleus, as well as on the speed of the expansion. Here the isentropic expansion of a nucleus is reviewed first. This exercise will identify the possible routes to a phase change namely, evaporation-condensation, liquid-to-gas phase transition, and isothermal and adiabatic spinodal decomposition. These possibilities will be examined in the rest of the chapter.

3.1 Isentropic Expansion

So far the existence of phases has been discussed, but not the processes leading to the formation of these. This section presents a brief description of the thermodynamics of expanding nuclear matter.

As seen in figures 2.1 and 2.4, both the isentropic and isothermal spinodals are inside the region where bubbles and droplets can coexist. A nuclear system expanding past these boundaries will respond with *negative compressibility* enlarging any small density fluctuation and disassembling the system into droplets.

An idealized collision between two large heavy nuclear ions could, in principle, form a compressed and heated system that might expand after the initial collision. [In terms of density, temperature and entropy, the medium would go from $(n, T, S) = (n_0, 0, 0)$ to some point with $n > n_0$, $T > 0$, and $S > 0$.] Since this hot and dense system is now isolated from the

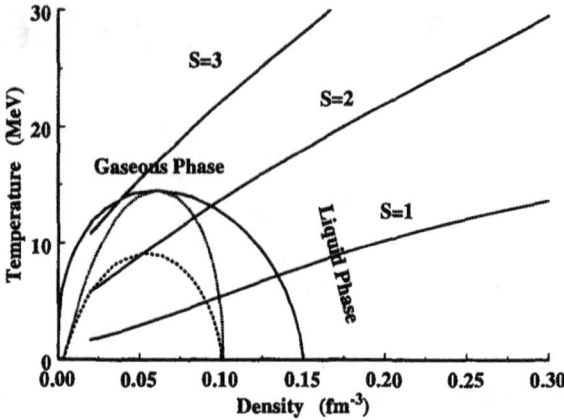

Fig. 3.1   Isentropic Expansion of Nuclear Matter

rest of the universe, it would expand adiabatically into vacuum conserving the total entropy generated during the impact. As the density of the blob drops, its temperature will decrease, but as long as the system remains in a single phase there will be no further changes of entropy. Its temperature evolution can thus be mapped to the n-T plane on a constant entropy path.

Since there is a one-to-one correspondence between any pair of points (n, T) and (n, S), one can solve for T from $p(n, T) = p(n, S)$, which, in our parametrization, is equivalent to $\varepsilon_T(n, T) = \varepsilon_S(n, S)$. The resulting expression is

$$T(n, S) = -T_B(n) + \frac{\sqrt{T_B^2(n) + 4\, T_A(n)\varepsilon_S(n, S)}}{2\, T_A(n)} \; , \qquad (3.1)$$

where, as in section 1.3, $T_A(n) = \sum_{i=0}^{2} \varepsilon_{i2}^T n^i$ and $T_B(n) = \sum_{i=0}^{2} \varepsilon_{i1}^T n^i$. Figure 3.1 shows the trajectories that expanding systems at $S = 1$, 2, and 3 would follow on the $n - T$ plane. Once the expanding nuclear system enters beyond the boundary of the coexistence region, it will decompose into liquid and gaseous phases. The final configuration will depend on the speed of the expansion as well as on the isentrope followed during the expansion.

Several possibilities emerge in the isentropic expansion of a nuclear system:

- **Isothermal Phase Transition**
  Gas-to-Liquid Phase Transition. Nuclear systems initially in the gaseous phase, and expanding from a starting point with high T or high S (*i.e.* above the critical point), as in the $S = 3$ curve in figure 3.1, could go into a mixed phase by condensation of gaseous particles into droplets in the superheated region.
  Liquid-to-Gas Phase Transition. Nuclei expanding slowly from an initial density in the liquid regime and subcritical T (or S), as in the $S = 1$ curve in figure 3.1, will have enough time in the supercooled region to allow the uniform liquid to nucleate forming a mixture of gaseous particles and droplets.
- **Spinodal Decomposition**
  Isothermal Spinodal Decomposition. If the expansion of a liquid system is fast enough, a nucleus starting from a liquid density and subcritical temperature or entropy (as in the $S = 2$ curve in figure 3.1) will not have enough time to nucleate in the supercooled region. In this case, the nucleus will be able to reach the isothermal spinodal and will break into a mixed phase by the isothermal growth of density fluctuations.
  Adiabatic Spinodal Decomposition. If the expansion of a nucleus is fast enough, and starts from a high density and low temperature or entropy (as in the $S = 1$ curve in figure 3.1), it will not disassemble in the supercooled region. This time, the nucleus will enter the isentropic spinodal and break into a mixed phase by spinodal decomposition.

These possibilities will be reviewed in turn in the following sections.

Problem 3.1   T(n, S)
Derive expression (3.1) starting from $p(n, T) = p(n, S)$ using the definitions introduced in sections 2.1 and 2.3.

3.2   Isothermal Phase Transition

A possible road towards the transformation of a saturated vapor to a mixture of liquid and vapor is via the growth of droplets embedded in the vapor. In the same way bubbles of vapor can appear in a liquid phase. Both the liquid-to-gas and the gas-to-liquid phase transitions take place in the metastability region, delimited by the coexistence curve and the isothermal spinodal (*cf.* figure 2.4). The analysis of such processes are described by nucleation theory.

Nucleation is that process in which a density fluctuation grows via the aggregation of nucleons or shrinks due to nucleon evaporation. Next, the conditions needed to have a stable phase mixture are analyzed in an increasing degree of complexity.

3.2.1 Nucleation: Bulk and Surface Effects

This section analyzes the conditions in which a spherical nucleus can coexist with a vapor of nucleons when only the surface tension of the nuclear drop is taken into account [Huang (1987)].

Consider a liquid nuclear drop surrounded by a gas of nucleons at pressure p. If the nucleus undergoes a small expansion dV, it will perform work described by $dW = pdV - \sigma da$, where da is the increase in surface area and $\sigma(T)$ the surface tension. The energy, E, and the Gibbs free energy, G, of the nucleus attain the forms, respectively,

$$E = \frac{4}{3}\pi r^3 \epsilon_\infty + 4\pi r^2 \sigma(T) \ , \quad G = \frac{4}{3}\pi r^3 g_\infty + 4\pi r^2 \sigma(T) \ ,$$

with ϵ_∞ and g_∞ representing the specific energy and Gibbs potential of the nucleus. [Remember that $g = \mu$ at constant temperature and pressure.]

Taking into account now the nucleons in the gaseous phase, the total Gibbs energy of this mixture is $G_t = M_g \mu_g + M_l \mu_l + 4\pi r^2 \sigma(T)$, where M_g, μ_g, M_l, and μ_l refer to the mass and chemical potential of the gaseous and liquid phases. If, out of this mixture, a drop of A nucleons is formed, the variation of G upon formation is

$$\Delta G_t = A(\mu_l - \mu_g) + 4\pi r^2 \sigma(T) \ . \tag{3.2}$$

Here the first term is what is called the *bulk* term, and the second one the *surface* term. The equilibrium radius can be found from $\partial(\Delta G)/\partial r = 0$. Using the liquid density as $n_l = A/(4\pi r^3/3)$ and the nuclear radius as $r = r_0 A^{1/3}$ in equation (3.2),

$$r_e = \frac{2\sigma(T)}{(\mu_l - \mu_g)n_l} \tag{3.3}$$

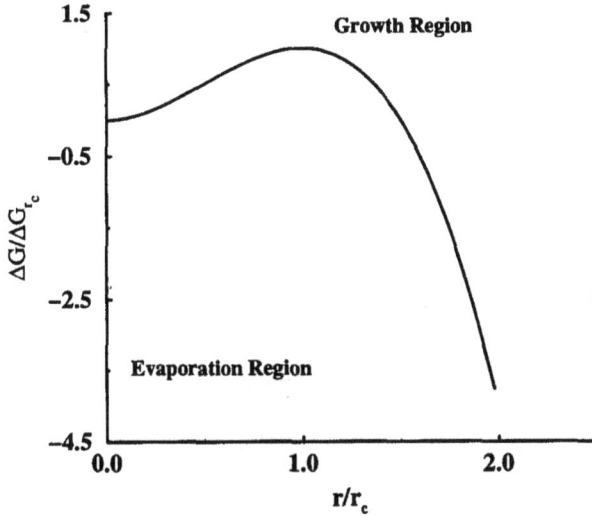

Fig. 3.2 ΔG as a function of r

which displays a maximum for $\mu_l < \mu_g$. Figure 3.2 shows the behavior of ΔG as a function of r for our *EOS*. Inserting equation (3.3) into (3.2), it is obtained that

$$\frac{\Delta G_t}{\Delta G_t]_{r=r_e}} = -2\left(\frac{r}{r_e}\right)^3 + 3\left(\frac{r}{r_e}\right)^2 .$$

The maximum of ΔG at r_e is the thermodynamic potential barrier. Because the Gibbs potential should be a minimum at equilibrium, drops with radius $r_d > r_e$ will grow absorbing particles from the vapor. On the other hand, drops with $r_d < r_e$ will evaporate particles and, eventually, disappear. Figure 3.3 shows the stability condition when $(\mu_l - \mu_g)$ is plotted as a function of r_e. Once again two regions are delimited, one in which drops grow and another in which drops evaporate.

It is also interesting to study the stability condition as a function of the pressure. Since $(\partial\mu/\partial p) = 1/n$, it can be written that

$$\frac{1}{n_g - n_l} = \frac{2\sigma(T)}{n_l r^2}\left(\frac{\partial r}{\partial n_l}\right) - \frac{2\sigma(T)}{n_l^2 r^2}\left(\frac{\partial n}{\partial p}\right) , \tag{3.4}$$

and the density of the vapor phase can now approximated as the one cor-

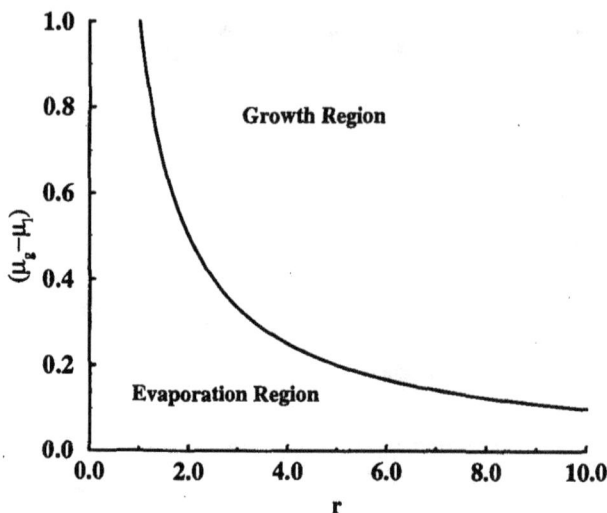

Fig. 3.3   Growth and evaporation regions

responding to an ideal gas, *i.e.* $n_g \approx mp/T$, where m is the mass in the volume under consideration. Moreover, since the liquid is much less compressible than the vapor, we approximate $\partial n_l / \partial p = 0$, and, finally, because $n_l \gg n_g$, then $1/n_l \ll 1/n_g$, from which equation (3.4) can be rewritten as:

$$\frac{T}{mp} = \frac{2\sigma(T)}{n_l r^2} \left(\frac{\partial r}{\partial p} \right) ,$$

from which it is obtained, after integration

$$p_r(T) = p_\infty(T) e^{(2m\sigma(T)/nTr)} ,$$

which gives the equilibrium pressure of a drop in a vapor as a function of the radius of the drop. This dependence is shown in figure 3.4, and one can immediately see that the higher the pressure the smaller the size of the equilibrated drop. This means that at high pressure a density fluctuation which involves only a few particles can trigger the transition from vapor to liquid.

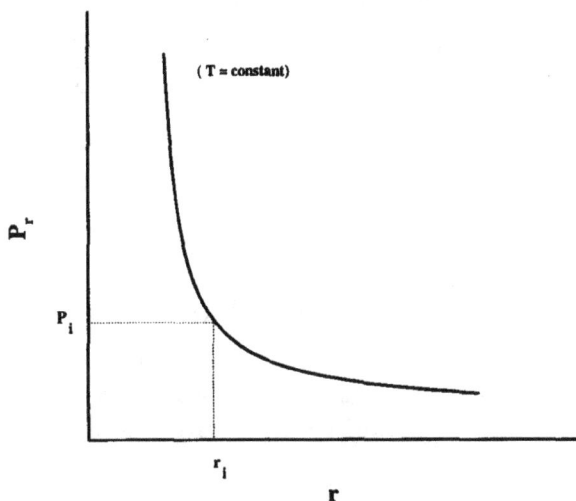

Fig. 3.4 Equilibrium pressure of a drop in vapor versus its radius. Higher pressures correspond to smaller drops, indicating that even small density fluctuations can trigger vapor to liquid changes at high $p_r(T)$.

3.2.2 Surface Curvature Correction

In the previous section, the only term related to the finiteness of the drop was the surface tension. But since this surface closes on itself, its entropy is restricted to those shapes that preserve the volume and size of the drop. This effect was considered by Fisher [Fisher (1971)] who proposed the following expression for the variation of the Gibbs free energy when a drop is formed:

$$\Delta G = A(\mu_l - \mu_g) + 4\pi r^2 \sigma(T) + \tau T \ln A .$$

In this equation the new effect is given by the last term, with the Fisher coefficient τ correcting the units.

Performing the same analysis as in the previous section, the variation of ΔG with A, can be calculated

$$\frac{d\Delta G}{dA} = (\mu_l - \mu_g) + \frac{2\sigma(T)}{n_l r} + \frac{\tau kT}{\frac{4}{3}\pi r^3 n_l} ,$$

from which the extremum condition can be solved to give

$$(\mu_g - \mu_l) = \frac{2\sigma(T)}{n_l r_e} + \frac{\tau kT}{\frac{4}{3}\pi r_e^3 n_l} = \frac{2\sigma(T)}{A_e^{1/3}(\frac{4}{3}\pi)^{1/3}(n_l)^{2/3}} + \frac{\tau kT}{A_e} \; . \; ,$$

where the subscript e refers to the equilibrium values. This cubic equation can be solved as done by [Goodman *et al.* (1984)]. Then, the region of drop growth (drop evaporation) is given by

$$[\mu_g - \mu_l] > (<) \left[\frac{2\sigma(T)}{n_l r_e} + \frac{\tau T}{\frac{4}{3}\pi r_e^3 n_l} \right] . \qquad (3.5)$$

Because both terms in the right hand side of equation (3.5) are bigger than zero, and both are decreasing functions of the radius r_e, the overall behavior of the system is not strongly altered by the curvature term.

3.2.3 *Charged Drops*

Although the Coulomb interaction is not explicitly included in our nuclear *EOS*, its effects are taken into account in the fitting of the thermodynamic variables (*e.g.* binding energy, *etc.*) at saturation densities. For completeness, the effect of the Coulomb interaction on the behavior of the liquid nuclear drops will now be explored.

Assuming that the nuclear drops are homogeneously charged with a charge Z equivalent to one half of the mass A (*i.e.* $Z = A/2 = \alpha A$, $\alpha = 1/2$), ΔG is now given by

$$\Delta G = A(\mu_l - \mu_g) + 4\pi r^2 \sigma(T) + \tau kT \ln A + \delta C \; ,$$

where δC stands for the contribution of the Coulomb energy. To analyze the effect of the new term, the variation in Coulomb energy must be studied during the formation of a drop in a region of charged vapor at constant vapor density. Notice that this implies a reduction of the overall volume of the system. The Coulomb energy of a charged drop in a homogeneously charged vapor by C it involves the following terms:

$$C = C_{dd} + C_{dv} + C_{vv} \; ,$$

with C_{dd} being the Coulomb energy due to the interaction between nucleons in the drop, C_{dv} the interactions between nucleons belonging to the drop and the rest of the system, and the last term, C_{vv} the interaction energy between particles in the vapor in the new reduced volume. The charge density of the liquid and vapor phases are given in terms of α: $n_l^c = \alpha e n_l$ and $n_g^c = \alpha e n_g$, with e denoting the unit of charge. In this way $\delta C = C - C^0$, with C^0 given by

$$C^0 = \frac{3}{5} \left(\frac{4\pi}{3}\right)^2 (\alpha e)^2 n_g^2 r_0^5 \, ,$$

i.e. the Coulomb energy of a charged vapor drop of radius r_0.

δC can be obtained in a simple approximation as the difference in Coulomb energy between a vapor drop of mass A and a liquid drop of the same mass [Bondorf *et al.* (1995)],

$$\delta C = \frac{3}{5} \left(\frac{4\pi}{3}\right)^2 (\alpha e)^2 (n_g^2 r_0^5 - n_l^2 r_l^5) \, ,$$

which can be expressed as:

$$\delta C = \frac{3}{5} (\alpha e)^2 \frac{A^2}{r_l} \left(1 - (n_g/n_l)^{1/3}\right) \, .$$

Calculating the derivative and taking into account that away from the critical point $n_l \gg n_g$,

$$\frac{d}{dA} \delta C = \frac{4\pi}{3} (\alpha e)^2 n_l r_l^2 \, . \tag{3.6}$$

It should be noticed that at the critical point $n_l = n_g$ and thus $\delta C = 0$.

Then, adding the right hand side of equation (3.6) to (3.5) the stability condition is obtained for the simple approximation of the Coulomb effects. For drop growth (evaporation) the condition is:

$$[\mu_g - \mu_l] > (<) \left[\frac{2\sigma(T)}{n_l r_e} + \frac{\tau k T}{\frac{4}{3} \pi r_e^3 n_l} + (\frac{4\pi}{3})(\alpha e)^2 n_l r_e^2\right]$$

It is immediate to see that the Coulomb term is an increasing function of the radius of the liquid drop.

The behavior of this stability condition is sketched in figure 3.5. It can be readily seen that the metastability line displays a minimum. This minimum determines a threshold below which no growth can take place; moreover, unlimited growth cannot take place. A drop formed above the metastability line will grow until this line is reached and further growth stops.

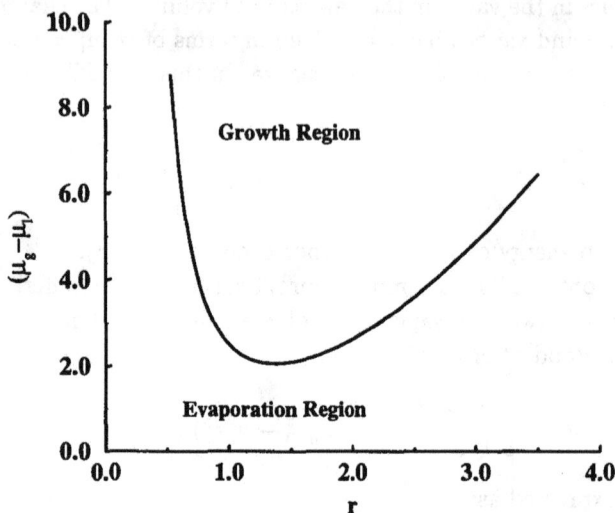

Fig. 3.5   The metastability line determines growth and non-growth regions.

Problem 3.2   The Coulomb Term

Show that the complete expression for the Coulomb term is

$$\delta C = C - C^0 = C_{dd} + C_{dv} + C_{vv} - C^0$$
$$= \left(\frac{4\pi\alpha e}{3}\right)^2 \left\{ r_l^3 (n_l - n_g) \left[\frac{3}{5} r_l^3 (n_l - \frac{3}{2} n_g) + \frac{3}{2} r_g^2 n_g \right] - \frac{3}{5} n_g^2 (r_0^5 - r_g^5) \right\} .$$

3.2.4   *Droplet Yield Distribution*

Fluctuation theory [Landau and Lifshitz (1980)] states that for systems in thermodynamic equilibrium, the probability, $w(x)$, that a magnitude x attains a value between x and dx is $w(x) \propto e^{S(x)}$, where $S(x)$ is the entropy of the system. Using the magnitude R_{min} (which characterizes the minimum work needed to vary the state of the system in a reversible way) it is obtained that $w(x) \propto e^{-R_{min}/T_0}$, with $R_{min} = \Delta E - T_0 \Delta S + p_0 \Delta V$, with ΔE, ΔS and ΔV being the energy, entropy and volume variations of the small region where a fluctuation takes place. T_0 and p_0 are the temperature and pressure mean values. In particular, when dealing with a system at constant temperature and pressure, $R_{\min} = \Delta G$.

In the case under analysis, the interest is on density fluctuations that give rise to drops. The probability of having a drop in the vapor is given by

$$P_r(A) \propto e^{-\Delta G/T} . \tag{3.7}$$

The expressions derived in section 3.2.1 can now be used to explore the different terms affecting the stability of liquid drops immersed in vapor.

Let us first consider the case in which ΔG is driven by bulk, surface and curvature terms. In this case equation (3.7) can be rewritten in the following way:

$$P_r(A) = Y_0 A^{-\tau} e^{-\left[(\mu_l - \mu_g)A + 4\pi r_0 \sigma(T) A^{2/3}\right]/T} ,$$

with Y_0 a normalization constant.

Functional Forms. Depending on the "position" on the coexistence curve, $P_r(A)$ will adopt different functional forms. Cases of special interest are analyzed now:

- **Supersaturated Region.** In the supersaturated region (segment CB in figure 2.2), $\mu_l < \mu_g$ and the yield is described by

$$P_r(A) = Y_0 A^{-\tau} e^{(\mu_g - \mu_l)A/T} e^{-\left(4\pi r_0^2 \sigma(T) A^{2/3}/T\right)} \tag{3.8}$$

  which displays a U-shaped behavior.
- **Coexistence Region.** If the system is on the coexistence region, then $(\mu_l - \mu_g) = 0$, *i.e.* the equilibrium condition, then

$$P_r(A) = Y_0 A^{-\tau} e^{-4\pi r_0^2 \sigma(T) A^{2/3}/T} . \tag{3.9}$$

  In this case a power law decay plus an exponential fall-off dominating for large masses is found.
- **Critical Point.** If the system is at the critical point, once again $(\mu_l - \mu_g) = 0$ but also $\sigma(T_c) = 0$, which states that liquid and vapor are indistinguishable at this point. The yield distribution is then

$$P_r(A) = Y_0 A^{-\tau} , \tag{3.10}$$

  which is a pure power law and, as such, free of scales.

In our approximation (see section 3.2.3), the Coulomb term that needs to be added to the variation of the Gibbs thermodynamic potential is $\delta C = 3(\alpha e A)^2 [1 - (n_g/n_l)^{1/3}]/(5r_l)$. As noticed before, this term vanishes at the critical point because $n_g = n_l$, this is another way of expressing that vapor and liquid are indistinguishable at the critical point. The yield is once again a pure power law. Away from the critical point, the Coulomb term is always present. It plays the same role as the one related to the surface energy but with a stronger dependence with the mass.

Surface tension. A relevant magnitude in this calculation is the surface tension $\sigma(T)$ [Domb and Green (1980)]. It is well known that $\sigma(T)$ vanishes at the critical point. It is generally assumed that it goes to zero as a power law of the form:

$$\sigma(T) \propto (T_c - T)^{\mu} \ ,$$

with μ being a critical exponent.

On the other hand if κ is the inverse of the surface thickness, it goes to zero at the critical point as

$$\kappa \propto (T_c - T)^{\nu} \ ,$$

with the critical exponent, ν, related to the divergence of the correlation distance at the critical point, *i.e.* $\zeta \sim (T_c - T)^{-\nu}$.

Since σ represents the excess free energy per unit of area and κ^{-1} is a length, then $\sigma\kappa$ is the excess free energy per unit volume in the interface

$$\sigma\kappa = (T_c - T)^{2-\alpha} \ ,$$

with α, the critical exponent, related to the divergence of the specific heat c_v at the critical point. From this $\mu + \nu = 2 - \alpha$ from which it can be written

$$\sigma(T) \propto (T_c - T)^{2-\alpha-\nu} \ .$$

Because of the general scaling relation $2 - \alpha = \nu d$ with the dimensionality, d, we get for a three dimensional system $2 - \alpha - \nu = 2\nu$. Then taking into account that the index ν for a classical fluid is about 0.62, it can be written

$$\sigma(T) = (T_c - T)^{5/4} \ .$$

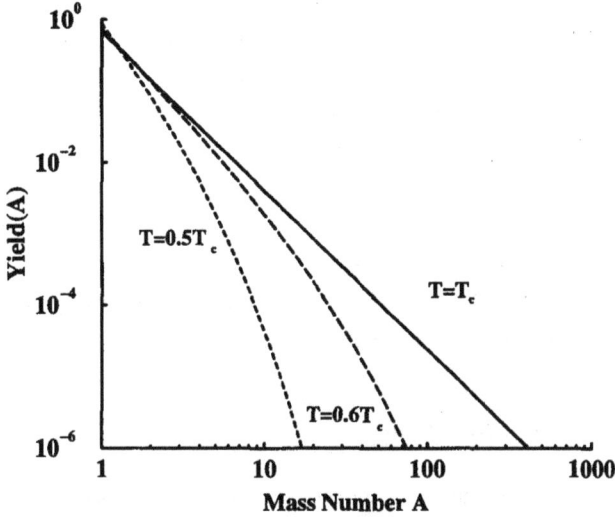

Fig. 3.6 Resulting mass yields versus the fragment mass for $T = 0.5\,T_c$, $0.8\,T_c$, and T_c.

A convenient interpolation formula that correctly reproduces the low temperature limit and the behavior around the critical point is

$$\sigma(T) = \sigma_0 \frac{\left(T_c^2 - T^2\right)^{5/4}}{\left(T_c^2 + T^2\right)^{5/4}} \; ,$$

with $\sigma_0 = 18\,MeV/4\pi r_0^2$. [Pethick and Ravenhall (1987)]

Yields. In order to have a qualitative view of the behavior of the yields, A can be fixed to $A = 200$ and use the critical temperature for our *EOS*, $T_c = 14.4\,MeV$. Considering the yield on the coexistence curve, then $\mu_l = \mu_g$, and no U-shaped distributions appear.

Figure 3.6 shows the resulting mass yields as a function of the mass of the fragments for $T = 0.5\,T_c$, $0.8\,T_c$, and $T = T_c$. No Coulomb term has been considered. It is readily seen that as we approach the critical point the yield goes from an exponential decay towards a pure power law.

Problem 3.3   Coulomb Correction

Analyze the effect of including the Coulomb correction in the calculation of the yield for the above mentioned temperatures. Notice that the liquid and gaseous densities are to be calculated. For this purpose use the condition of equal areas in the Maxwell construction.

Problem 3.4   Yield Out of Coexistence Region

Calculate the yield when the system is out of the coexistence curve. In this case $\mu_l \neq \mu_g$. Calculate the values of the corresponding chemical potentials by integrating $V dp$ along the chosen isotherm.

3.3   Spinodal Decomposition

Before an expanding nuclear drop enters the spinodal, it will spend some time in the metastable region. If the expansion is slow enough, the system might be long enough in this region to allow for the nucleation process take place. Nucleation, however is a long process and, in fast-expanding nuclei, the system might enter the spinodal and disassembly by the growth of density fluctuations. This section reviews the kinematics and time scales of the spinodal decomposition.

3.3.1   *Kinetics of Spinodal Decomposition*

The first step in the study of the kinetics of spinodal decomposition is to consider the temporal behavior of small density fluctuations δn via the diffusion equation as first done by [Cahn (1961); Hillert (1961); Pethick and Ravenhall (1987)].

$$\frac{\partial n}{\partial t} = \nabla \cdot (M \nabla \mu) ,$$

where M is the mobility of the medium. In this case the force driving the diffusion comes from $\nabla \mu$, *i.e.* from the difference in chemical potential between the disturbance and the surrounding medium. The particularization of this equation to the case of small density fluctuations comes through the relation between μ and the Helmholts free energy.

The free energy density of an inhomogeneous medium can be expressed as a sum of the free energy of a uniform system and the one associated with the generation of incipient surfaces by the fluctuations, *i.e.*

$$f_i = f(n) + B(\nabla n)^2/2 \,,$$

where B is the van der Waals constant. Writing μ in terms of the free energy, the diffusion equation can be generalized (after linearizing in n) to

$$\frac{\partial n}{\partial t} = M\frac{\partial^2 f}{\partial n^2}\nabla^2 n - MB\nabla^4 n \,.$$

Describing now an arbitrary fluctuation in terms of its Fourier components, the general solution of this equation for each of these components is an oscillatory function of the form $\delta n = A(\vec{q}, t)\cos(\vec{q}\cdot\vec{r})$, with the amplitude given by

$$A(\vec{q}, t) = A(\vec{q}, 0)e^{-Mq^2(\partial^2 f/\partial n^2 + Bq^2)t} \,.$$

Two kinematics regions are readily seen. If $\partial^2 f/\partial n^2 + Bq^2 > 0$, the fluctuations are damped away as normally expected, but if $\partial^2 f/\partial n^2 + Bq^2 < 0$ the inhomogeneities will be amplified as time goes on, and the medium will be unstable with respect to these oscillations. This instability is termed spinodal decomposition, as it is the detonator for the end of an homogeneous phase.

The speed of sound c_s can be related to the previous condition using $c_s^2 = M(\partial^2 f/\partial n^2)$, to yield $\exp\left[-(c_s^2 + MBq^2)q^2 t\right]$ for the amplitude exponential. [The second term, MBq^2, can be taken as a q-dependent term of the speed of sound responsible for the anomalous dispersion of sound.] The kinetic amplification factor, or growth rate squared, $\Gamma^2 = -c_s^2 q^2 - MBq^4$, shows different magnitudes for different wavenumbers.

The behavior of Γ^2 can be studied for our *EOS* as a function of q^2 using c_s^2 from expression (2.2), and the nuclear value of $B = 80\, MeV\, fm^5$ [Pethick and Ravenhall (1987)]. For temperature and density combinations lying inside the isothermal spinodal, $c_s^2 < 0$ and $\Gamma^2 > 0$ indicating the growth of the fluctuations and passage to a mixture of phases. Figure 3.7 shows the magnitude of the growth rate squared in the middle of the spinodal region as a function of q for several temperatures. At this density, the growth vanishes for $T > T_c$, *i.e.* Γ^2 becomes negative indicating a damped oscillatory response from the nuclear medium.

The wavenumbers for which the nuclear system is unstable, are certainly of relevance for real nuclei. For instance, the $q = 1\, fm^{-1}$ mode (which is well inside the growth region for $0 < T < 10\, MeV$) corresponds to an

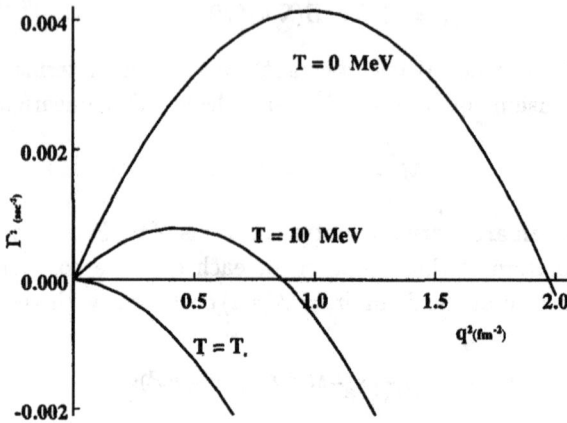

Fig. 3.7   Growth rate squared of density fluctuations

oscillation of wavelength $\lambda = 2\pi/q \approx 6.2 \ fm$. This perturbation is of the right size to make a 100-particle nucleus with a radius of $\approx 5.6 \ fm$, decay. For a study of the time evolution of density fluctuations entering the spinodal region, see [López and Lübeck (1989)].

Problem 3.5   Maximum Unstable Wavenumber
Determine the maximum wavenumber q_{max}^2 at which unstable modes exist inside the spinodal. Use expression (2.2) to sketch q_{max}^2 as a function of T for $n = 0.15 \ fm$. Hint: $\Gamma^2(q_{max}^2) = 0$.

Problem 3.6   Maximum Growth Rate of Density Fluctuations
Determine, Γ_{max}^2, i.e. the maximum growth rate squared of density fluctuations inside the isothermal spinodal, and sketch it as a function of T for $n = 0.15 \ fm$. Hint: Γ_{max} occurs at the top of the q^2-parabola, i.e. at $q_c^2 = q_{max}^2/2$, (cf. problem 3.5). Use expression (2.2) for the speed of sound.

3.3.2   *Other Issues in Spinodal Decomposition*

Time Scales. The analysis of the kinematics of spinodal decomposition presented in section 3.3.1 is valid for both fluctuations developing isothermally or isentropically. The time scales for these two processes, however, are different. For adiabatic fluctuations to develop inside the spinodal, the disassembly process must be shorter than the time needed for the energy to be transported over the lenght of the nucleus. The characteristic time for energy transport is given by the thermal diffusion equation.

Isentropic Expansion. As explained in section 3.1, and proposed by [López and Siemens (1984)], an expanding heavy nucleus follows an isentropic path down to the isentropic spinodal region and then disassembles by spinodal decomposition. During the separation of phases the system jumps from the initial isentropic path onto an isotherm curve. The final isotherms can be determined by introducing the gas mass fraction α, and requiring that the density n and energy density E of the post-fragmentation state satisfy

$$1/n = \alpha \frac{1}{n_g} + (1 - \alpha)\frac{1}{n_l},$$
$$E = \alpha E_g + (1 - \alpha)E_l,$$

where n_l, n_g, E_l and E_g are the values for the liquid and gaseous phases at the phase transition. The final isotherm is then the one that satisfies these two equations, and it can be obtained by tabulating the n_l, n_g, E_l and E_g for different temperatures. The final entropy is also found by means of $S = \alpha S_g + (1 - \alpha)S_l$.

3.4   As a Conclusion

The simple equation of state of the chapter 2 is complete enough to analyze the transition from a uniform phase to a mixed phase under a number of conditions. For instance, parametrization of $T - n$ trajectories of systems in isentropic expansions showed the possibility of a *gas-to-liquid* phase transition when the expansion starts at high T or S, or to a *liquid-to-gas* phase transition for slow expansions staring from high densities and small T or S. Other decay possibilities are the *isothermal* or *adiabatic spinodal decompositions* for fast expanding nuclei starting from high density and subcritical temperature or entropy.

The model thus developed also allowed a direct application of nucleation theory to determine the phases composition after the phase changes. The conditions for phase coexistence were determined as well as the nuclear metastability line. Disassemblies taking place in the *supersaturated* region were found to display mass yields with a U-shaped form. Those occurring in the coexistence region produce yield with a power-law decay shape plus an exponential fall-off. The special case of phase changes taking place at the critical point produce a pure power law yield, free of scales.

Using the nuclear equation of state with Cahn's diffusion equation theory allowed a study of the kinetics of spinodal decomposition. Two kinematics regions were easily spotted, one where the density fluctuations are damped away, and another one where the inhomogeneities are amplified breaking the medium, *i.e.* the spinodal region. The speed of sound, which was also found to detect this region, was used to quantify the rate of growth of these instabilities. Studying the maximum unstable wavenumbers, the phenomenon of spinodal decomposition was found to be very relevant for the fragmentation of large nuclei.

Chapter 4

Critical Phenomena in Finite Systems

4.1 Introduction

So far we have been concerned with the thermodynamic description of liquid-gas phase transitions in infinite nuclear systems. Using statistical thermodynamics, nucleation processes have been observed to display mass distributions that evolve from a "U" shape to an exponential decay as the temperature of the system is raised. Somewhere in between this two extremes, there is a temperature at which the distribution of the sizes, s, is a pure power law with exponent τ. At this peculiar temperature, called the critical temperature T_c, the fragment size distribution (FSD) is proportional to $s^{-\tau}$. Such a distribution is free of scales (see problem 4.1) and, thus, fluctuations of all sizes are to be expected. The behavior of the system around T_c is known as critical behavior.

The finding of experimental nuclear mass spectra that could be fitted by a power law in proton-induced fragmentation of Krypton and Xenon targets [Hirsch $et\ al.$ (1984)] triggered the interest on the study of critical phenomena in nuclei. * The two main lines of research are the calculation of critical exponents of nuclear matter, and the determination of the caloric curve of nuclei. Despite the enormous efforts during the last decade, the determination of the nuclear critical exponents and the caloric curve continues to be an active area of research.

A number of complications absent in infinite systems, appear in the nuclear case. For one thing, finite size effects are introduced by the fact

---

*It was latter shown that the value of the observed critical exponent was due to the mixing of different events classes and integration over impact parameter.

that nuclei are composed by just a few tens of particles. Likewise, since excited nuclear systems are produced by means of nuclear reactions, these are formed out of equilibrium. Moreover Coulomb instabilities come into play and modify possible critical behavior.

The effect of the system size on a liquid-to-gas transition is very pronounced and difficult to quantify. For instance, when enough energy is added to a cold liquid drop, it begins to evaporate particles. At high energies, however, the drop undergoes a fragmentation process, breaking into many pieces and expanding. This collective radial motion is characteristic of finite systems and has no counterpart in the infinite case. The connection between this breakup of an excited drop and a phase transition is not well understood yet.

This chapter reviews the main ideas of criticality in infinite systems, and explores the theoretical tools needed to connect them to finite systems. [For completeness, appendix A presents some non-linear aspects of the fragmentation of small systems using the Lyapunov exponents.] Since the nuclear equation of state does not include a solid phase, the focus will be on "liquid-like" to "vapor-like" transformations. For this purpose, two simple models will be introduced, namely percolation and molecular dynamics.

The first one, percolation, is a purely geometrical model with true critical behavior that allows the extraction of critical exponents from fragment size distributions. The second one, classical molecular dynamics, helps in understanding the role played by dynamical effects in finite fragmenting systems. In particular, the time scales and dynamics of fragment formation, as well as the collective radial motion and its connection to temperature and excitation energy (*i.e.* the caloric curve) will be studied in this chapter.

4.2 Critical Phenomena in a Nutshell

Critical phenomena [Stanley (1971); Goldenfeld (1992); Yeomans (1994); Binney (1995)] refers to the behavior of matter around the critical temperature (if in a second order phase transition) or the critical point (in a first order one). First order phase transitions are characterized by the presence of latent heat, which is absent in second order ones. Second order phase transitions are usually related to variations in the symmetry of the system under study, (*i.e.* when crossing the critical temperature a new symmetry appears in the system) although there are exceptions to this rule.

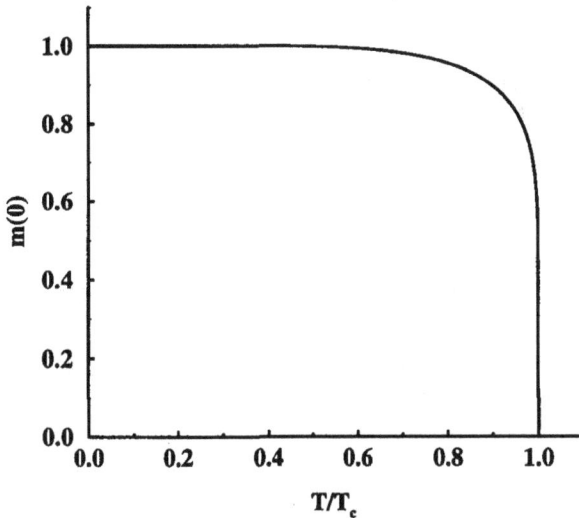

Fig. 4.1   Magnetization per spin for a two-dimensional Ising system in the absence of external magnetic field

[Typical examples of symmetry breaking and non-breaking second order phase transitions are the ferromagnetic-paramagnetic transition in Fe, and the gas to liquid transition in water, respectively. When the magnetization m of Fe is measured slightly above $T_c = 1044 \, K°$ (Curie point) and without an external magnetic field B applied, we find that $m(B = 0)$ vanishes and the system is isotropic. On the other hand, when the system is cooled below T_c, $m(0)$ becomes finite and a symmetry axis appears in the system, see figure 4.1. A case without a symmetry change is the transition from water vapor to liquid at the critical point. Since it does not involve latent heat, it is continuous and does not change the symmetry properties of water.]

Order parameters. The different properties of a system above and below the critical temperature are quantified by a quantity known as the order parameter. In general, order parameters are quantities with a vanishing thermal average on one side of the transition, but non-zero on the other. For example, this parameter is the magnetization $m(0)$ in paramagnetic-ferromagnetic transitions, whereas for the vapor to liquid change it is the difference between the gas and liquid densities, $(\rho_g - \rho_l)$, which equals zero at and above the critical point, but not below (see section 3.2.4). Order parameters have been determined for a number of phenomena including critical opalescence in binary fluids, and the $He_I - He_{II}$ transition, among others.

Critical exponents. An interesting feature of the critical point is that in its vicinity the behavior of the systems is marked by the fact that various thermodynamic quantities posses singularities. It is usual to express this singularities in terms of power laws. The powers of these relationships (six, traditionally denoted by α, β, γ, δ, η and ν) are called "critical exponents", which determine the qualitative nature of the critical behavior of the system under study. The following table summarizes their definitions:

$$
\begin{aligned}
c &\sim |T - T_c|^{-\alpha} & &\text{specific heat,} \\
m(0) &\sim |T_c - T|^{\beta} & &\text{magnetization at } B = 0, \ (T < T_c) , \\
\chi &\sim |T - T_c|^{-\gamma} & &\text{susceptibility ,} \\
(m)_{T=T_c} &\sim B^{1/\delta} & &\text{magnetization at } B \neq 0 , \\
G(r) &\sim 1/r^{d-2+\eta} & &\text{pair correlation function at } T_c , \\
\xi &\sim |T - T_c|^{-\nu} & &\text{correlation length (valid where } G(r) \sim e^{-r/\xi}).
\end{aligned}
$$

Please notice that in the previous table we have refered to magnetic systems. If one is interested in gas-liquid transitions one should replace m by $(\rho_l - \rho_c)$ or $|\rho_l - \rho_c|$ and B by $(P - Pc)$

Universality classes. When dissimilar systems have similar critical exponents, they are said to belong to a "universality class." For example, a power law of the form $\sim |T - T_c|^{-\gamma}$ is satisfied by the susceptibility or iron (with $\gamma = 1.3$) and the isothermal compressibility of water ($\gamma = 1.2$). Several classes have been determined to exist (via the renormalization group [Stanley (1971); Goldenfeld (1992); Yeomans (1994); Binney (1995)]), as for example the so-called three-dimensional Ising model which comprises fluids, ferromagnetic materials, binary alloys, etc. In this way, critical exponents are the "fingerprints" of a phase transition: to characterize a critical phenomenon of an unknown type its corresponding critical exponent must be compared with the corresponding ones of several universality classes.

Scaling hypothesis. The classification of critical phenomena can be further reduced by four relationships among the critical exponents resulting from the "Scaling Hypothesis" of [Widom (1965)] and [Kadanoff (1966)]. The fact that the order parameters are discontinuous and derivable from the free energy indicates that the energy is a singular function. Separating the free energy per particle in the vicinity of the critical point as a sum of a regular part and a singular one, i.e. $f(T, B) = f_r(T, B) + f_s(T, B)$, the scaling law proposes a specific dependence of f_s on T and B.

Letting ϵ represent $|T - T_c|$, the singular part of the free energy is assumed to depend, not on T and B separately but, on the ratio $B/\epsilon^{x/y}$, with x and y being parameters to be determined. Mathematically,

$$f_s(\epsilon, B) = \epsilon^{1/y}\psi(B/\epsilon^{x/y}) , \tag{4.1}$$

with ψ being a function of a single variable. Obtaining the zero-field magnetization through $m_{B=0} = -(\partial f/\partial B)_T$, and since close to T_c, $m \sim \epsilon^\beta$, the first critical exponent becomes $\beta = (1 - x)/y$. Using similar arguments for the susceptibility, specific heat, and other order parameters, it can be obtained that $\gamma = (2x - 1)/y$, $\alpha = 2 - 1/y$, and $\delta = x/(1 - x)$. See problem 4.2.

These results for β, γ, σ and δ show that these critical exponents are not independent. Kadanoff also obtained similar relationships for η and ν. In fact only two of the critical exponents are independent, thus, eliminating the redundant parameters, the relations among them are:

$$\begin{array}{lll} \text{Fisher} & \gamma = \nu(2 - \eta) & \\ \text{Rushbrooke} & \alpha + 2\beta + \gamma = 2 & \\ \text{Widom} & \gamma = \beta(\delta - 1) & \\ \text{Josephson} & d\nu = 2 - \alpha . & \end{array} \tag{4.2}$$

Where d stands for the dimensionality of the systems under study. See problem 4.3.

Problem 4.1 Scale-Free Power Laws
Power laws such as $g(s) = s^{-\tau}$ are free of scales because, given the right amplification, a sketch of the function around s could be superimposed on a sketch at, say, $100s$. Calculate the growth ratio of $g(s)$ over a decibel centered at $10s$ (i.e. $g(10s)/g(s)$) and compare to that at $100s$. Is this scaling possible with other dimensionless functions such as $h(s) = e^s$?

Problem 4.2 Scaling Hypothesis
A. *Use the zero-field magnetization $m_{B=0} = -(\partial f/\partial B)_T$ with expression (4.1) to show that the first critical exponent is $\beta = (1 - x)/y$.*
B. *Use the susceptibility $\chi_{B=0} = -(\partial^2 f/\partial B^2)_T$ to obtain $\gamma = (2x - 1)/y$.*
C. *Use for the zero field specific heat $c_B = -T(\partial^2 f/\partial B^2)_T$ to obtain $\alpha = 2 - 1/y$. Hint: let $B \to 0$ to eliminate all derivatives of ψ.*

Problem 4.3   Critical Exponents

Use the four scaling expressions for the critical exponents, namely
$\beta = (1 - x)/y$, $\gamma = (2x - 1)/y$, $\alpha = 2 - 1/y$, and $\delta = x/(1 - x)$ *to eliminate x and y and obtain the Rushbrooke and Widom relationships of equations (4.2).*

4.3   Finite Systems

In the preceding section, the main results of the critical phenomena in infinite systems were presented. In recent years a growing interest on the extension of those ideas to finite systems has taken place. Here two of such ideas, namely finite size scaling theory, and an analysis of the "solid-like" to "liquid-like" phase transition in small clusters will be reviewed.

4.3.1   *Finite Size Scaling*

As mentioned in the previous section, in finite systems all divergences are replaced by maxima in positions shifted with respect to the infinite case. The main question arising is, how to describe the behavior of finite systems in the neighborhood of the critical temperature? In this case, the scaling form of the singular part of the free energy density takes the form [Goldenfeld (1992)]:

$$f_s(\epsilon, L^{-1}) = \epsilon^{2-\alpha} f_s(\epsilon^{-1/\nu_*}, L^{-1}) , \qquad (4.3)$$

which can be written in terms of $\xi_\infty(\epsilon)$, *i.e.* the bulk correlation length corresponding to the infinite system $(L = \infty)$,

$$f_s(\epsilon, L^{-1}) = \epsilon^{2-\alpha} f_s(\xi_\infty L^{-1}) .$$

When $L \gg \xi_\infty(\epsilon)$ the correlation length of the system is not affected by the boundaries of the system, and the thermodynamic properties are those of the infinite system. On the other hand, when the system approaches the critical point, the correlation length goes to infinity and the boundaries of the system play a noticeable role. For example, from equation (4.3), the specific heat has the form:

$$c(\epsilon, L^{-1}) = L^{\alpha/\nu} D(\epsilon L^{1/\nu}) ,$$

where $D(x)$ is a scaling function with a maximum at x_0. Thus, the specific heat will have a peak at a temperature shifted from the one corresponding to the infinite system: $\epsilon_L = x_0/L^{1/\nu} \propto L^{-1/\nu}$ or $T = T_c + x_0/L^{1/\nu}$.

4.3.2 Solid-like to Liquid-like Transitions in Small Systems

The discovery of a transition to liquid-like behavior in computer simulations of atomic clusters of as little as seven constituents, has fueled a great interest in this kind of behavior and its connection with the thermodynamic limit. The main line of analysis focuses on obtaining the caloric curve of small systems.

In particular, [Labastie and Whetten (1990)] have shown, calculating the classical state density $\Omega(E)$ for clusters of 13, 55 and 147 Lennard-Jones particles, that the caloric curves clearly display three regions. In the first one, corresponding to a solid state, the excitation energy E increases steadily as a function of T and gives a constant specific heat. The second one is a transition region where the specific heat peaks in a single sharp increase. And finally the third region corresponds to a liquid state in which the slope of E is again low giving, with an almost constant specific heat. These calculations, which were performed in an energy range where evaporation is irrelevant, also show that, as the size of the cluster is increased, the maximum of the specific heat sharpens and shifts towards the value corresponding to an infinite system.

This clearly shows that the solid-like to liquid-like phase transition in finite systems resembles the behavior of the infinite case limit. To further clarify this point, figure 4.2 shows the caloric curve corresponding to a cluster of 55 Lennard-Jones particles obtained by molecular dynamics. It should be noticed [Labastie and Whetten (1990)], however, that there is no spatial coexistence in finite systems, but rather only a temporal coexistence, in which the system "jumps" back and forth between solid-like to liquid-like configurations.

4.4 Percolation Model

In what follows the properties of small systems undergoing a fragmentation will be investigated. First an analysis of the fragment size distribution resulting from a percolation in small lattices will be presented along with

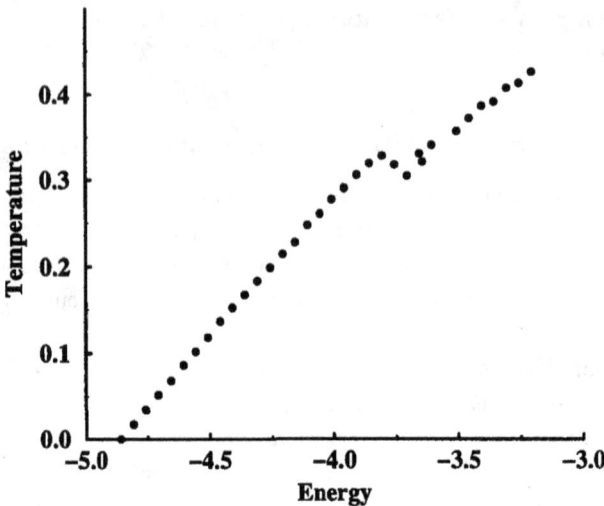

Fig. 4.2   Temperature of a three dimensional 55 particles Lennard-Jones drop as a function of the energy. The "loop" signals the solid-like to liquid-like transition

a nuclear percolation model.

Percolation [Stauffer and Aharony (1992)] is the most simple example of a system that displays critical behavior. Site percolation can be described as a grid of dimension d in which the sites are randomly populated with some probability (p). If instead of the nodes the internode links are activated with some probability (q), the process is known as bond percolation. And finally, if both nodes and bonds participate in the process, it is known as site-node percolation. Every bond percolation problem can be reduced to a node one, but the converse is not true. All three percolation types are purely geometrical.

In a model like this, the phase transition (or the critical point) is related to the appearance of a percolating cluster in the system, *i.e.* a set of connected bonds or occupied sites going from $-\infty$ to $+\infty$. Several definitions of a percolating cluster exist for finite systems. One such definition, for instance is the one in which occupied neighboring sites (activated bonds) extend from one side of the system to the opposite one. For infinite systems there is a sharp critical site occupation probability p_c such that for $p(q)$ above p_c, the probability of finding a percolating cluster is 1, whereas below

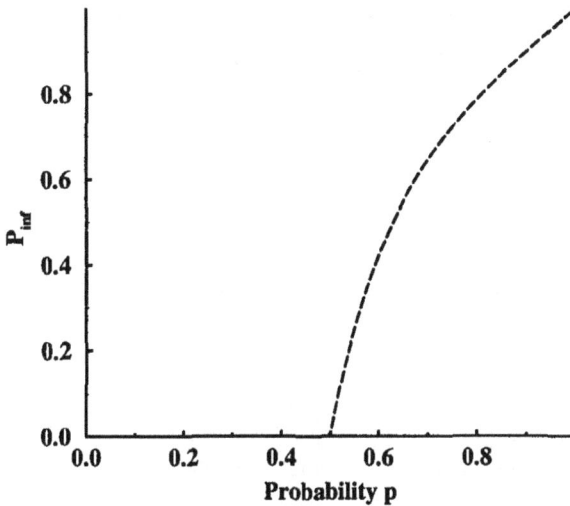

Fig. 4.3 Strength of the maximum fragment for the Bethe lattice, P_∞, as a function of the concentration, or occupation probability p.

p_c the probability is strictly 0. There is a similar critical bond activation probability q_c for bond percolation. For finite lattices the transition from one regime to the other is smooth, i.e the probability of finding a percolating cluster is different from 0 for any probability.

In what follows we make only reference to the site problem in order to ease the notation. In this case this second order phase transition is characterized by the order parameter given by the fraction of sites that belong to the percolating cluster, P_∞. In figure 4.3 we can see that P_∞ displays the standard behavior for order parameters around the critical point, it is *zero* below p_c and different from *zero* above. The behavior of P_∞, for $p > p_c$ and $p \sim p_c$, is then given by $P_\infty \sim (p - p_c)^\beta$

Now, taking into account that the probability of a site being empty is $(1 - p)$, that of a site belonging to the infinite cluster is $p \cdot P_\infty$, and that of a site belonging to a finite cluster is $\sum_s s \cdot n_s$ (with n_s being the number of clusters of size s), we can write the following sum rule:

$$(1 - p) + p \cdot P_\infty + \sum_s s \cdot n_s = 1 .$$

The critical properties of percolation are vividly represented by the singular behavior of the moments of the fragment size distribution:

$$m_0 = \sum_s n_s(p) \sim |p - p_c|^{2-\alpha}$$

$$m_1 = \sum_s s \cdot n_s(p) \sim (p - p_c)^\beta$$

$$m_2 = \sum_s s^2 \cdot n_s(p) \sim |p - p_c|^{-\gamma}$$

$$m_k = \sum_s s^k \cdot n_s(p) \sim |p - p_c|^{(\tau-1-k)/\sigma}$$

and the correlation length

$$\xi(p) \sim |p - p_c|^{-\nu} \ .$$

As stated in § 4.2, the critical exponents are related among them, in this case these relationships are:

$$\gamma = (3 - \tau)/\sigma \qquad \beta = (\tau - 2)/\sigma$$
$$\alpha = 2 - (\tau - 1)/\sigma \qquad D = 1/(\tau - 1) \ ,$$

where D is the fractal dimension of the percolating cluster at p_c. The numerical values of these quantities are determined, not by a phenomenological scaling theory, but from microscopic methods. For infinite 3-dimensional cubic lattice, the values of the critical exponents are: $\tau = 2.18$, $\beta = 0.41$, $\gamma = 1.8$ and $\sigma = 0.45$ [Stauffer and Aharony (1992)].

Near the critical point the typical cluster size, s_ξ, becomes very large. The size s_ξ can be defined as the one that gives the main contribution to the sums in the singular part of the moments. In accordance with the behavior of the correlation length ξ, the typical cluster size diverges at the critical point. Following Stauffer, we further assume that there is a unique typical cluster size.

It seems plausible then that the singular properties of the clusters should depend on the single ratio s/s_ξ. Thus the relevant quantity regarding cluster sizes should be

$$n_s/n_{s_\xi} = F(s/s_\xi) \ ,$$

a quantity independent of the distance $p - p_c$. This relationship is referred to as a scaling assumption on cluster sizes. Noticing that for $p \sim p_c$, n_s behaves as $s^{-\tau}$, Stauffer proposed a generalization of the droplet model [Fisher

(1971)],

$$n_s(p) \sim s^{-\tau} f(z) , \qquad (4.4)$$

where the scaling variable is $z = (p - p_c)s^{\sigma}$, and the scaling function is $f(z)$. It is known that $f(z)$ has only one maximum located below the critical point, and that $f(0) = 1$.

When we are dealing with finite systems we must resort to finite size scaling. In the case of the strength of the maximum fragment, it takes the following form:

$$P_\infty^{(L)}(p) = L^{-\beta/\nu} \overline{P}(L/\xi) ,$$

and if $x = (L/\xi)$, then $\overline{P}(x \gg 1) \sim L^{\beta/\nu}(p - p_c)^{\beta}$ in order to recover the right asymptotic behavior.

4.4.1 *Percolation of Small Lattices*

The results on percolation presented in the previous section refer to infinite systems and to the theory which relates the limiting behavior of finite lattices. Since for nuclear systems the number of lattice sites cannot exceed a couple of hundreds, the focus now shifts to small lattices to study their behavior as a function of their sizes. All the results shown in this section are obtained using the program Percol.for described in appendix B.4.

Fragment Size Distribution. First the behavior of the fragment size distribution (FSD) will be investigated. As already stated in section 4.1 the FSD is of the form:

$$n_s(q) \sim s^{-\tau} f(z) . \qquad (4.5)$$

Figures 4.4 and 4.5 show the FSD as a function of the bond probability q for a $5 \times 5 \times 5$ and a $30 \times 30 \times 30$ simple cubic lattices, respectively. It should be noticed that for the $5 \times 5 \times 5$ system approximately 70% of the particles belong to the surface, while for the $30 \times 30 \times 30$ case, only about 19% of the particles are on the surface of the system. The finiteness of the system is apparent in the behavior of the FSD for big fragments. Comparing figure 4.4 with figure 4.5, one can readily see that, although the expected transition from "U shape" to power law, and to an exponential decay can be seen in the smaller lattice, it is much more noticeable on the bigger system.

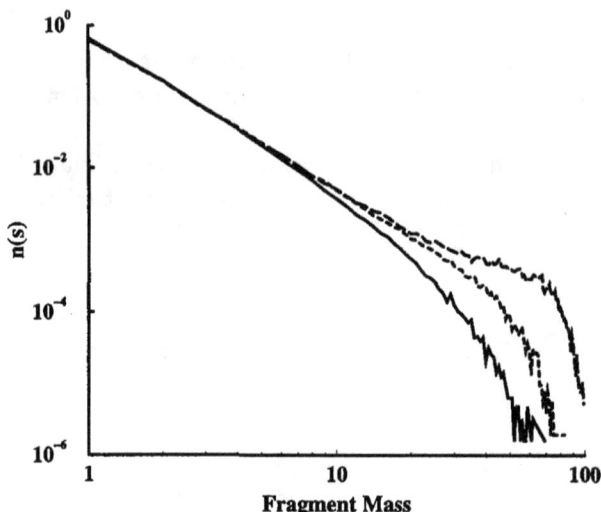

Fig. 4.4   Fragment size distribution for a 5 × 5 × 5 simple cubic lattice with bond activation probabilities of $q = 0.2$ (full lines), $q = 0.25$ (dashes), and $q = 0.3$ (long dashes). The effect of finiteness of the system is quite noticeable on the bigger fragments of the high-q curve.

Since at the critical bond probability, the *FSD* attains a pure power law shape, $n_s(q) \sim s^{-\tau}$, and at $z = 0$ the scaling function is $f(z = 0) = 1$, it is then tempting to try a power law fit to gain extra knowledge about the system.

Assuming then that, in the region of intermediate mass fragments, $n_s(q) = a_0 s^{-\tau_e}$ with a_0 being a normalization constant, the value of q for which the minimum value of τ_e is obtained, can be associated to q_c, and τ_e to τ. However, the finite size of the system introduces an additional complication.

From (4.5) it is easy to see that our assumption leads to

$$b_0 s^{-\tau} f(z) = a_0 s^{-\tau_e} \, ,$$

with b_0 also a normalization constant. The minimum of τ_e, can then be obtained by solving $f'(z) = 0$. However, since the maximum of $f(z)$ is shifted from 0, the resulting value of q will be shifted from the true critical point. Therefore, this simple method of determining q_c is not universal and

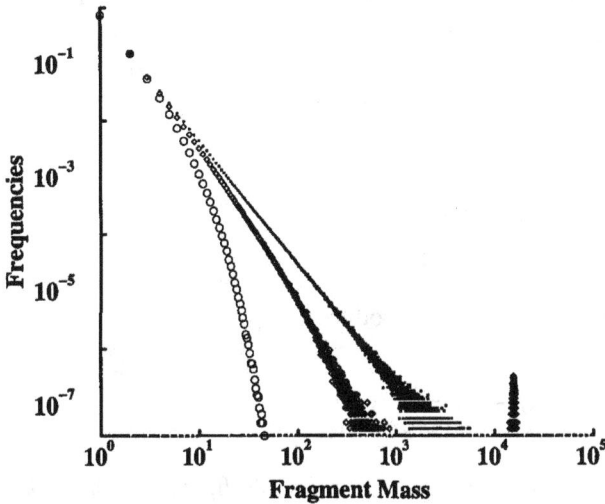

Fig. 4.5 Fragment size distribution for a $30 \times 30 \times 30$ simple cubic lattice for three values of the bond activation probability, $q = 0.2$ (circles), $q = 0.25$ (diamonds) and $q = 0.3$. (squares). The presence of a big fragment at $q = 0.3$ gives rise to a "U" shape distribution.

should be refined. Another possible way to extract reliable information about the critical exponents from finite-systems data is from the analysis of the moments of the fragment distribution.

Fluctuations and Averages. The basic quantity from which all properties of a fragmentation spectra can be obtained, is the probability of observing a given partition of the system described by the set of numbers $\{n_1, n_2, n_3, n_4, \ldots\}$ with n_i the multiplicity of fragments of size i. Denoting such a probability as $P\{n_i\}$, then an approximation to it can be obtained by analyzing a large number of experiments and getting the observed frequency of the set of numbers $\{n_i\}$:

$$f\{n_i\} = \frac{N\{n_i\}}{\sum_{\{n_i\}} N\{n_i\}} ,$$

and $N\{n_i\}$ being the number of events in which the specific configuration $\{n_i\}$ was detected. Notice that the total number of experiments is $\sum_{\{n_i\}} N\{n_i\} = N_t$, and in the limit $N_t \to \infty$, we get $f\{n_i\} \to P\{n_i\}$.

However such a detailed information about the fragmentation process is usually not needed, and/or not so easy to handle. Notice that for a cubic lattice of size L in d dimensions, $P\{n_i\}$ is defined in a space of L^d dimensions. It is more convenient to deal with less detailed information like, for example, the probability of finding the number n_s in the set $\{n_s\}$. Such a projection can be easily obtained noticing that the mean value of, say n_i, is given by

$$\langle n_i \rangle = \frac{\sum_{\{n_i\}} n_i N\{n_i\}}{\sum_{\{n_i\}} N\{n_i\}} \,,$$

which, in the limit of ∞ experiments, produces

$$p(n_i) = \frac{\langle n_i \rangle}{\sum_{1 \leq j \leq A} \langle n_j \rangle} \,,$$

with A the total mass of the system. For example, figure 4.4 shows the fragment frequencies for a lattice of $5 \times 5 \times 5$ sites as a function of the bond probability q.

In addition to these mean values, every value of s has a fluctuation around the average, and it is possible that there is some hidden information on these fluctuations. Given an ensemble of M_0 stochastic variables (*i.e.* M_0 throws of a perfect coin), then the ratio of the standard deviation to the average value, $\sigma_s / < n_s >$, goes to zero as $1/\sqrt{M_0}$, if the fluctuations are purely statistical. Any departure from this behavior can contain interesting information about the problem under analysis. It should be kept in mind that for processes with other sources of fluctuations, for example in M_0, the analysis becomes much more complicated. [Campi (1986); Campi and Krivine (1986)]

A good way to reduce the sources of fluctuations is to specify a subset of the data, for example fixing M_0. Labeling these classes by α, the mean fragment size distributions can now be further classified as $< n_s, \alpha >$, with α denoting, for instance, multiplicity, size of the biggest fragment, etc. In what follows, all averages will be over sets that belong to the same class, for example, with the same multiplicity.

The data coming from a given fragmentation experiment (assuming that all fragments are recorded), or from a fragment recognition analysis of a numerical calculation can be presented in the following form:

$$
\begin{array}{ccccccccc}
1 & \to & n_1^1 & n_2^1 & \dots & n_f^1 & \Rightarrow & m_k^1 & \kappa_j^1 \\
2 & \to & n_1^2 & n_2^2 & \dots & n_f^2 & \Rightarrow & m_k^2 & \kappa_j^2 \\
\dots & \dots & \dots & \dots & \dots & \dots & \dots & \dots & \dots \\
l & \to & n_1^l & n_2^l & \dots & n_f^l & \Rightarrow & m_k^l & \kappa_j^l \\
& & \Downarrow & \Downarrow & \dots & \Downarrow & & \Downarrow & \Downarrow \\
& & & & & & & <m_k> & <\kappa_j> \\
& & <n_1> & <n_2> & \dots & <n_f> & \Rightarrow & \bar{m}_k & \bar{\kappa}_j
\end{array}
$$

The first column denotes the fragmentation event. The values n_j^i denote the multiplicity of fragments of size j in event i. And the symbols κ_j^i denote other quantities depending on the event by event multiplicities, moments, etc., as for example γ_2. Given the data organized in this way we can perform two kinds of averaging processes, an "horizontal" one and a "vertical" one.

For the "horizontal" case, for each event we can calculate a set of moments which we denote as m_k. These moments are defined as before: $m_k = \sum_{s=1}^S s^k n_s$. As stated above, m_0 is the multiplicity, m_1 is related to mean values via $\bar{s} = m_1/m_0$, and m_2 appears in $\sigma_s^2 = (m_2/m_0) - (m_1/m_0)^2$, which is closely related to the magnitude $\gamma_2 = m_2 m_0/m_1^2$ [Campi (1986); Campi and Krivine (1986)].

It is interesting to notice that for "U" shape distributions (*i.e.* those including a big fragment), the presence of the large fragment will dominate the behavior of the moments with $k \geq 2$, rendering them insensitive to the rest of the distribution. It is then obvious that, if all fragments are recognized in each event, as with numerical simulations, m_1 will contain no information. This suggest that it is better to remove the biggest fragment from n_s letting $n_s \Longrightarrow n_s - \delta(s - s_{\max})$.

From this analysis the total average is

$$
<m_k> = \frac{1}{N} \sum_j m_k^j. = \frac{1}{N} \sum_j \sum_{s=1}^S s^k (n_s - \delta(s - s_{\max})) .
$$

On the other hand, vertical analysis yields $<n_i>$, from which the vertical mean values $\bar{m}_k$ can be obtained:

$$
\bar{m}_k = \sum_{s=1}^S s^k \frac{1}{N} \sum_j (n_s - \delta(s - s_{\max})) .
$$

If the variables under analysis are linear in n_s, and taking into account that the events belong to the same class, for instance with the same multiplicity, both analyses will give the same results. In particular for a linear case, it is easy to show from these two last equations that $\overline{m}_k = < m_k >$. When the objects under analysis are not linear in n_s, like for example in γ_2, the results will differ, giving information about the fluctuations.

Problem 4.4 $\overline{m}_k$ and $< m_k >$
Show that if the s^k are linear in n_s, belonging to the same multiplicity class, then $\overline{m}_k = < m_k >$.

Moment Analysis This section presents an analysis of "fragmentation data" obtained with the program `Percol.f` of appendix B.4. Using this program, bond percolation events for different 3-dimensional simple cubic lattice sizes were generated. Each event is characterized by a random value of the bond probability. Nuclear fragmentation events were "simulated" without any discrimination of events by means of, for example, impact parameter. In this way, no information about the energy deposited in the system is available, and thus all kind of events are to be expected. This is the reason why in this simplified approach q is left as a free random parameter.

As stated in previous sections, the moments of percolation are defined by

$$m_k = \sum_s s^k \cdot n_s(p) \sim |p - p_c|^{(\tau - 1 - k)/\sigma} = \varepsilon^{\mu_k} ,$$

where $\sim$ stands for "the singular part behaves as". It must be noticed that the calculation of the moments requires knowledge about the distance from the critical point, $|p - p_c|$, but it must be kept in mind that it is assumed that this information is not available.

Using the normalized moments defined by $m'_k = m_k/m_1$, this difficulty can be avoided by noticing that [Campi and Krivine (1994); Belkacem *et al.* (1995)]:

$$\log(m'_k) = \log(m'_{k+n}) \frac{(\tau - 1 - k)/\sigma}{(\tau - 1 - k - n)/\sigma} .$$

Figure 4.6 shows a plot of m'_2 versus m'_3 in a double logarithmic scale, with each point corresponding to one event from the percolation simulation. Because this quantities do not diverge in finite systems but display a maximum around q_c, it is possible to extract the critical exponents graphically from figure 4.6.

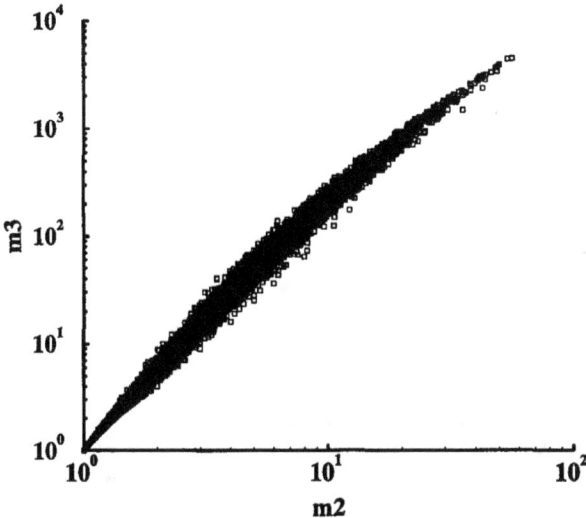

Fig. 4.6   The logarithm of m_3 as a function of the logarithm m_2. Event by event analysis over 4000 percolation experiments in a $6 \times 6 \times 6$ simple cubic lattice.

According to this last equation, the slope should be $\mu = (\tau - 4)/(\tau - 3)$. From the slope of the highest values of m_3 and m_2 in figure 4.6, we get $\mu = 2.22$, which compares well with $\mu = 2.3$ for infinite 3-dimensional percolation.

Another interesting way of getting information about critical exponents comes from the correlation between m_L, the size of the maximum fragment, and m_2'. This is shown in figures 4.7 (with 4,000 events considered), and 4.8 (with 50,000 events). Two branches can be observed, the upper one with a very small slope is hard to be analyzed, while the other one is such that

$$\log m_L = (1 + \frac{\beta}{\gamma}) \log m_2' .$$

Here it was used the fact that, while for an infinite system the critical size of the maximum fragment goes like $P \sim s^{\beta}$, the size of a typical cluster in a finite system behaves as $m_L(\varepsilon) \sim \varepsilon^{-(\gamma+\beta)}$.

The fluctuations increase very strongly with the decrease of the number of events taken into account. Figure 4.8 indicates that $1 + \beta/\gamma \cong 1.3$. For infinite-system percolation $1 + \beta/\gamma \simeq 1.23$ and for a liquid-gas phase

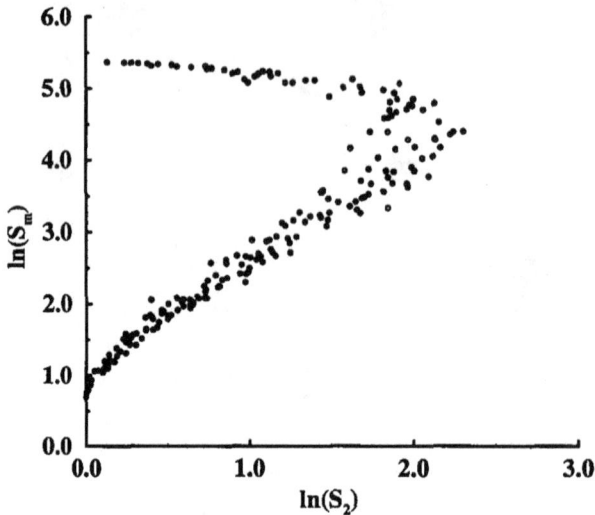

Fig. 4.7 Campi plot. Logarithm of the maximum fragment size against the logarithm of m_2, averaged over 4,000 events belonging to the class of same multiplicity, for a simple cubic lattice of $6 \times 6 \times 6$ with random activation bond probabilities between $0 < q < 1$.

transition $1 + \beta/\gamma \cong 1.5$.

The critical exponents obtained from these calculations are not close enough to the percolation values as to determine the Universality class the system should belong to. In references [Campi and Krivine (1994)] and [Belkacem *et al.* (1995)] this analysis was extended to experimental data and, although the values obtained are consistent with those of the previous calculation, it is still not possible to characterize the nuclear case. Nevertheless, the occurrence of values consistent either with percolation or liquid-gas transition strongly suggests that a critical phenomena is taking place.

Another quantity of interest is the normalized variance of the size of the maximum fragment (NVM) given by

$$NVM = \frac{\sigma_{A_{max}}}{< A_{max} >} .$$

Figure 4.9 shows the values of NVM obtained for lattices of $5 \times 5 \times 5$ and $30 \times 30 \times 30$ [Dorso *et al.* (1999)].

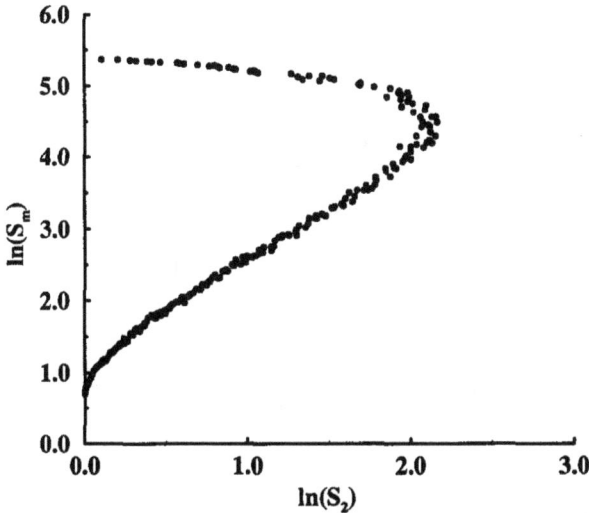

Fig. 4.8   Same Campi plot as in the previous figure but for 50,000 events.

The percolation data indicates that NVM has a maximum at the effective critical q, and that this maximum gets sharper and shifted towards the infinite size limit as the dimension of the lattice is increased. This measure of the fluctuations is quite robust against "unphysical"(finite size effects) fluctuations.

Two methods for the extraction of critical exponents in lattices as small as 6^3, *i.e.* of nuclear size, has been discussed recently [Elliot *et al.* (1997)]. One rests on the determination of $f(z)$ from which the critical exponents are determined, while the other one determines the exponents through a series of assumptions and fits.

To implement the latter method, it was realized that the vanishing of the divergence deforms the power behavior or m_2 in the vicinity of p_c, and that for large values of $(p - p_c)$ the relations (4.4) are no longer valid. The procedure devised excludes the biggest cluster S_{max} from the sums in m_k. Calling the region of $p > p_c$ the liquid region and $p < p_c$ the gas region, the relation for m_2 (*cf.* equation (4.4)) is supposed to be valid on each region. Assuming that γ attains a unique value above and below p_c as in infinite systems, p_c can then be determined by varying it until the power law fits

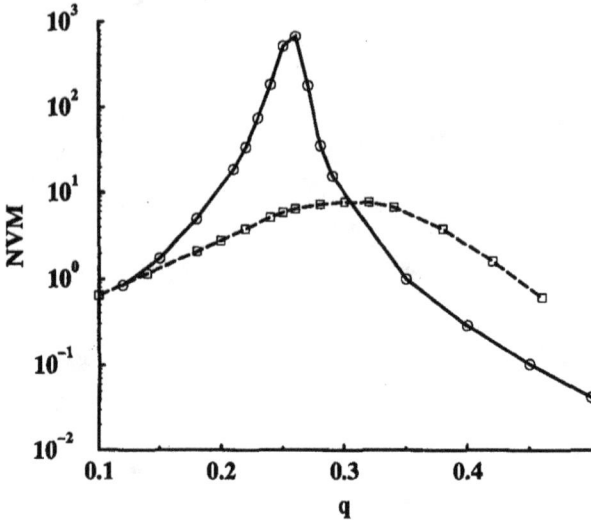

Fig. 4.9   Normalized variance of the maximum fragment size as a function of the bond probability q. For a $5 \times 5 \times 5$ simple cubic lattice (dashed lines) and for a $30 \times 30 \times 30$ simple cubic lattice (full line)

give the same value of γ in each region. The second critical exponent, from which the rest can be deduced, is obtained from $S_{max} \sim |p - p_c|^{\beta}$.

Table 4.1 shows the values obtained for different lattice sizes and compares them to physical systems. The columns $L = \infty$, $L = 63$ and $L = 6$ list results for infinite size percolation, a big finite lattice, and a nuclear-like one, respectively. These values are compared to those of a standard liquid-gas phase transition and Au fragmentation (obtained by the *EOS* collaboration [Gilkes *et al.* (1994)]).

The values listed in table 4.1 for the $L = 63$ lattice are very good, but those for a "nuclear" size lattice (*e.g.* $L = 6$) show the effects of the finite size. The value of β turns out to be the most difficult to determine, whereas the exponent τ is rather insensitive to the system size.

Problem 4.5   Percolation

The percolation computer program listed in appendix B.4 provides fragment sizes and multiplicities. Use this code to explore the following:

- **(a)** *Choose a lattice dimension and a set of bind activation proba-*

Table 4.1   Critical exponents

Exponent	$L = \infty$	$L = 63$	$L = 6$	liquid-gas	Au
τ	2.18	2.21 ± 0.02	2.2 ± 0.1	2.21	2.14 ± 0.06
β	0.41	0.46 ± 0.02	0.2 ± 0.3	0.33	0.29 ± 0.02
γ	1.8	1.87 ± 0.02	1.73 ± 0.007	1.23	1.4 ± 0.1
ν	0.88			0.63	
σ	0.45		0.5 ± 0.04	0.64	

bility q. Build the corresponding *FSD*, calculate the effective τ_{eff} and its minimum value, and identify the effective critical value of q for each dimension. Study the dependence of q on the dimension of the lattice.

- **(b)** *Using the same set of events as in (a) calculate the magnitude γ_2 with and without making a class selection of the events. Study this magnitude as a function of q and size.*

- **(c)** *Taking S as the total mass of the system, explore the behavior of $\log(m_i/S)$ versus $\log(m_j/S)$ for different sets of $\{i, j\}$ and discuss the possibility of extracting information of the critical exponents in each case.*

- **(d)** *Compare the behavior of $\log(m_2/S)$ versus $\log(m_0/S)$ for a small lattice (i.e. $5 \times 5 \times 5$) and for a big one ($30 \times 30 \times 30$). Evaluate in which region could it be possible to extract information about the big system from the small one.*

- **(e)** *The same as in (d) but using the mean value of the maximum fragment as a function of $\langle m_0 \rangle$*

4.4.2   *The Nuclear Lattice Model*

The simple geometrical model of percolation was adapted to study nuclear fragmentation using bond percolation [Bauer (1998)]. In this model the nuclear excited system is represented by a three dimensional lattice of a given topology, with the nodes representing "nucleons". The only parameter (apart from size and topology of the grid) is the probability p of a bond being broken. This probability can be related to the amount of energy deposited in the system.

Different recipes have been developed to accomplish this mapping. In one such approach [Bauer (1998)], the nucleus under consideration is represented by a spherical system with a breaking probability p given by

$p = E^*/E_B$, with E^* representing the excitation energy of the nucleus and E_B its binding energy. E_B can be written as $E_B = (z/2)E_{bond}$, with z being the number of closest neighbors. Notice that $E^* \leq E_B$ in order to have $p \leq 1$.

It is further assumed that the energy distributed in each bond of the systems, ϵ_b, can be properly described by a Boltzmann distribution with mean energy $< \epsilon_b >$ [Li *et al.* (1994); Phair *et al.* (1993)]. If α denotes the mean number of bonds associated at each site, then the average energy deposited on each site is $< E_s >= \alpha < \epsilon_b >$, and the initial binding energy of the nuclear system is $B = \alpha E_b$. Then, the bond breaking probability is given by

$$p = \frac{\int_{E_b}^{\infty} \sqrt{\epsilon_b} e^{-\epsilon_b/t_b}}{\int_0^{100} \sqrt{\epsilon_b} e^{-\epsilon_b/t_b}} = \frac{\int_B^{\infty} \sqrt{E_s} e^{-E_s/T_s}}{\int_0^{100} \sqrt{E_s} e^{-E_s/T_s}},$$

with $t_b = (2/3) < \epsilon_b >$, and $T_s = \alpha t_b = 2E_s/3$ as the slope parameters. T_s is determined by fitting proton kinetic energy spectra with a single moving Boltzmann source.

Other formulations of the nuclear lattice model include approaches using site-bond percolation [Campi and Desbois (1985); Desbois (1987)] and a grid-free percolation [Dorso *et al.* (1990)].

4.4.3 *Caveats*

A drawback of using percolation on nuclear problems is that it is a purely geometrical model without any genuine dynamical effects, such as the radial collective motion. To explore dynamical effects another computational model is needed. Next, the method of molecular dynamics will be applied to the study of small drops. This technique solves the equations of motion of all particles of the system, thus producing all the dynamical microscopic information.

4.5 Molecular Dynamics

Even though the percolation model of the previous section captures some relevant aspects of fragmenting finite systems, it lacks the dynamical ingredients. To investigate dynamical aspects in disassembling finite systems, the computational method known as Molecular Dynamics (*MD*) will be used. This method uses classical mechanics with interactions dictated by a two-body potential to study the dynamics of a collection of particles.

Although realistic implementations of nuclear MD models exist [Pandharipande *et al.* (1990); Barrañón *et al.* (1999)], it is pedagogically beneficial to use a simpler two-dimensional model with a well known potential. In this section, the process of nuclear fragmentation will be simulated studying the time evolution of excited $2D$ Lennard-Jones disks. The computer program used can be downloaded by anonymous ftp from zapata.utep.edu or from the author's web page at www.utep.edu/physics.

The properties of the Lennard-Jones potential are very well known. Its repulsive part at short distances combines with an attractive part at larger distances to produce liquid and gaseous phases. The Lennard-Jones interaction potential is given by

$$V(r) = \begin{cases} 4\epsilon \left[\left(\frac{\sigma}{r}\right)^{12} - \left(\frac{\sigma}{r}\right)^6 - \left(\frac{\sigma}{r_c}\right)^{12} + \left(\frac{\sigma}{r_c}\right)^6 \right] & r < r_c \\ 0 & r \geq r_c \end{cases} \tag{4.6}$$

where r_c is the cutoff radius (taken as $r_c = 3\sigma$), and σ is the length scale constant. The units of time and energy are $t_0 = \sqrt{\sigma^2 m / 48\epsilon}$ and ϵ, respectively.

4.5.1 *Caloric Curve of Excited Drops*

In macroscopic systems phase transitions can be identified using the caloric curve (*cf.* § 1.5.1). This curve is defined as the functional relationship between the temperature of a system and the energy deposited in it. In the fragmentation problem isolated drops evolve in a constant pressure environment, in particular $p = 0$, thus the infinite system of reference should be an isobaric one. In general, the relation between temperature and energy for the homogeneous regions should behave as

$$\Delta T = \frac{\Delta E}{c_p} .$$

Figure 4.10 shows the caloric curve of water at constant pressure. The "rise-plateau-rise" pattern in which the "rising" regions are straight lines (one each for the solid, liquid and the vapor branches) is typical of infinite systems.

To use molecular dynamics to explore the caloric curve, it is necessary to

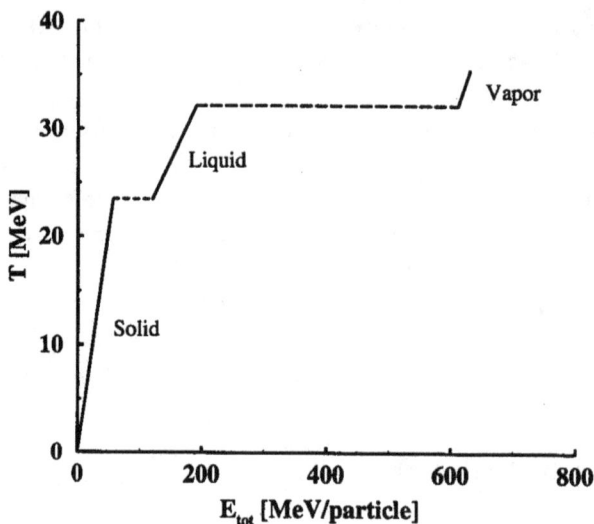

Fig. 4.10   Caloric curve of water at a pressure of 1 atm.

define its meaning for disassembling finite systems. Effectively, this caloric curve is the functional relationship between the temperature of the finite system at fragmentation time and the energy deposited in it. But to obtain this curve from MD simulations, it is necessary to know how to identify the fragmentation time, and how to measure the temperature from the fragments. This is specially difficult to do in a finite system. The next two sections present a chronology of cluster recognition algorithms and their use in extracting the caloric curve from MD simulations. Also discussed are the extraction of temperatures and how to relate the simulations with actual nuclear fragmentation data.

4.5.2   *Fragment Recognition*

In order to calculate the fragmentation times we need to be able to recognize fragments from the dynamical data resulting from our simulations. The first step is to perform the MD simulation of the breakup of $2D$ Lennard Jones disks. Once the trajectories of all the particles are known, it is necessary to pinpoint the formation of clusters. Modern fragment recognition techniques have evolved from configuration methods (MST), and now take into account relative momentum ($MSTE$) and binding energy ($ECFM$).

MST. The most widely used algorithm for recognizing fragments is the "Minimum Spanning Tree" (MST), which has been used in astronomy and other problems in configuration space, like percolation. In this approach a cluster is defined in the following way: given a set of particles $i, j, k, ...$, they belong to a cluster C if:

$$\nabla \; i \in C \;,\exists \; j \in C \; / \; |\vec{r}_i - \vec{r}_j| \leq R_{cl} \;,$$

where $\vec{r}_i$ and $\vec{r}_j$ denote the positions of the particles, and R_{cl} is a parameter usually referred to as the clusterization radius. This is an arbitrary parameter fixed according to taste. Usually the inter-particle potentials used in microscopic simulations are truncated at a cut-off distance R_{co}, so that $R_{cl} \leq R_{co}$.

In the definition of MST clusters, the role of the relative momentum of the particles is totally disregarded. In this way, two particles that are close in coordinate space will be recognized as forming a cluster even though they can be flying away due to a large relative momentum. It should then be clear that this method can only be used to analyze asymptotic configurations in which the fragmenting system is a dilute mixture of free particles and fragments almost in equilibrium. Consequently, if MST is used to describe the dynamical evolution of the fragmentation process (*i.e.* not the asymptotic stage) the results thus obtained will be wrong.

MSTE. An improvement over the MST algorithm, although in the same spirit, is to look for simply-connected structures in the space of two particle binding energy [Pratt *et al.* (1995)]. This model, denoted as $MSTE$, is defined in the following way: if i, j denote particles and C a cluster, then a cluster is a set of particles such that:

$$\nabla \; i \in C \;,\exists \; j \in C \; / \; e_{ij} \leq 0 \;,$$

with $e_{ij} = V_{ij} + (\vec{p}_i - \vec{p}_j)^2/4\mu$, where μ is the reduced mass. In this way, a cluster is built out of bound pairs of particles. It is clear that $MSTE$ takes into account the relative momentum between pairs of particles, but since a collection of bound pairs does not guarantee a bound cluster, the method is not perfect.

ECFM. A more evolved cluster recognition algorithm is the "Early Cluster Formation Model" ($ECFM$) introduced by [Strachan and Dorso (1997)]. According to $ECFM$, the clusters formed in a disassembling systems are those that produce the most bound partition of the system. That is, the selected partition (*i.e.* the set of clusters $\{C_i\}$) minimizes the sum of the energies of each fragment

$$E_{\{C_i\}} = \sum_i \left[\sum_{j \in C_i} K_j^{cm} + \sum_{j,k \in C_i} V_{j,k} \right] \, ,$$

where the first sum is over the clusters of the partition, and the second one over the constituent particles in each cluster. K_j^{cm} is the kinetic energy of particle j measured in the center of mass frame of the cluster which contains particle j.

It can be seen that this definition contains a recursive element, the binding energy cannot be calculated until the partition (and its center of mass) is defined, and the partition cannot be identified until the binding energy is known. Trying to solve this riddle by a "brute-force" analysis of every possible partition of the system is impossible due to the large number of possibilities. This problem can be simplified with the use Monte Carlo techniques.

ECRA. The coupling of *ECFM* and simulated-annealing like techniques gave birth to the "Early Fragment Recognition Algorithm" or *ECRA*. [Strachan and Dorso (1997)] In this approach, Monte Carlo techniques are used to sample possible partitions while searching for the lowest binding energies. When the method is applied to the early part of the disassembly process, the identified clusters are not the asymptotic fragments, but only the "most bound density fluctuations" (*MBDF*). These density fluctuations will become fragments in the low density regime, when they will coincide with *MST* fragments. Figure 4.11 shows the time evolution, according to the *MST*, *MSTE* and *ECRA* algorithms, of the size of the maximum fragment detected during the breakup of a highly excited 2*D* disk of Lennard-Jones particles.

The figure immediately shows that *ECRA* not only recognizes the maximum fragment earlier than *MSTE* and *MST*, but also indicates that it forms very early in the evolution. Without the *ECRA* information, *MST* would lead us to believe that the asymptotic maximum fragment is produced by a smooth evaporation process. This means that fragments are actually formed in this early stage, and then carry information about the early, dense and hot stage of the fragmentation process. In this case, power law behavior of the asymptotic mass spectra is obtained for $E = -0.6\epsilon$.

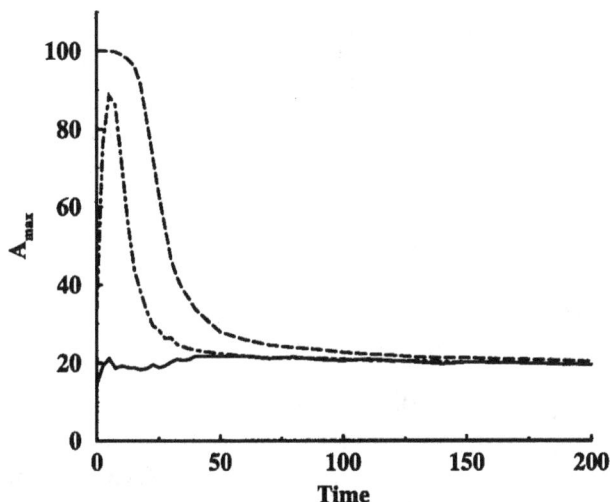

Fig. 4.11 Size of biggest cluster of MD fragmentation of 100 particle disks with $\epsilon = 0.4\epsilon_0$ with time in units of t_0 (see text). Fragment recognition algorithm used are: MST (long dashes), $MSTE$ (dot-dashes) and $ECRA$ (full line). $ECRA$ reveals that the $MBDF$ associated with the maximum fragment is present since very early.

4.5.3 *The Caloric Curve*

With the possibility of using $ECRA$ to identify the most bound density fluctuations at any time during the reaction, their study should allow now for the calculation of the caloric curve. The procedure includes studying the MD-generated evolutions with two of the fragment recognition algorithms described before, MST and $ECRA$. And once the fragments have been recognized, their microscopic stability is to be analyzed (using the microscopic variable known as "persistency") and its temperature determined. [Strachan and Dorso (1999)]

Time Scales, Fragment Stability and Persistence. Two different time scales can be defined, a "fragment formation time", τ_{ff}, related to the formation of the $ECRA$ partition, and a "fragment emission time" τ_{fe} corresponding to the MST partition. The $ECRA$ partition evolves into the MST one by a simple evaporation-like process. These times can be identified by studying the stability of the fragments, which can be done with a microscopic coefficient named "persistence".

At a given time t the system will be formed by a set of clusters $C_i(t)$ which will eventually become the asymptotic fragments, denoted by C_i. Considering a given cluster $C_i(t)$ with mass number $n_i(t)$, the number of pairs of particles in the cluster is $b_i(t) = n_i(t)(n_i(t)-1)/2$. Since at asymptotic times the constituent particles might be in two or more different clusters, it is convenient to track $a_i(t)$ as the number of pairs of particles that belong to $C_i(t)$ and are also together in a given asymptotic cluster. The persistence coefficient is thus defined as

$$P(t) = \frac{1}{N_{ev}} \sum_{ev} \frac{1}{\sum_{cl} n_i(t)} \sum_{cl} n_i(t) \frac{a_i(t)}{b_i(t)} \ ,$$

where the first sum runs over the different events for a given energy, N_{ev} is the number of events, and the other two sums run over the clusters existing at time t. It is clear that the persistence coefficient is equal to 1 if the microscopic structure of the clusters is equal to the asymptotic one. On the other hand, this coefficient approaches 0 when the two partitions under study bear little similarity. This coefficient can be defined for $ECRA$ clusters as well as for MST ones.

Figure 4.12 shows a typical result of $P(t)$ for the $ECRA$, MST and $MSTE$ partitions. In each case the asymptotic partition was taken as that resulting from the corresponding analysis ($ECRA$, MST or $MSTE$). Asymptotically $ECRA$ and MST yield almost the same results, as expected. The horizontal lines represent a reference value related to an evaporation-like process on the asymptotic configuration. It is the value of the persistence coefficient when each asymptotic partition is compared to one obtained by removing just one particle from each asymptotic cluster.

Again the $ECRA$ results reveal that the $MBDF$ attain microscopic stability before the corresponding structures are recognized by the other algorithms. Also of interest is that at τ_{ff} the biggest MST cluster still contains more than half of the total mass of the system, indicating that at fragmentation time the system is still highly interacting.

Temperatures. Once the appropriate time scales have been defined, the next step is to calculate the temperature at these times. Care must be exercised in calculating the temperature in a disassembling system, as this develops a radial collective motion. Simulations show that, in agreement with hydrodynamics, the velocity of the radial flow is position dependent growing linearly away from the center of the system.

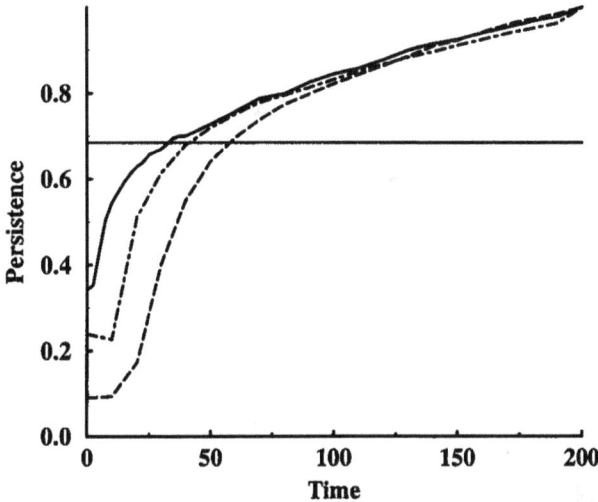

Fig. 4.12 The persistence coefficient for the same numerical simulation as the previous figure. For MST analysis (long dashes), for $MSTE$ (dot dashes) and for $ECRA$ (full line).

For this reason it is convenient to divide the $2D$ expanding drops in concentric circular regions and introduce local temperatures on these rings. Placing an origin at the center of mass of the system, and dividing the system into rings of, say, width $\delta r = 2\sigma$, the mean radial velocity of region i can be defined as

$$v_{rad}^{(i)}(t) = \frac{1}{N_i(t)} \sum_{ev} \sum_{j \in i} \frac{\vec{v}_j(t) \cdot \vec{r}_j(t)}{|\vec{r}_j(t)|} \, ,$$

where the first sum runs over the different events for a given energy, the second over the particles j that belong to region i at time t, $\vec{v}_j$ and $\vec{r}_j$ are the velocity and position of particle j. $N_i(t)$ is total number of particles belonging to region i in all the events.

With this, and assuming that the fragmenting system is in local equilibrium, the local temperature can be defined as

$$T_{loc}^{(i)} = \frac{2}{3} \frac{1}{N_i} \sum_{j \in i} \frac{1}{2} m \left(\vec{v}_j - \frac{v_{rad}^{(i)} \cdot \vec{r}_j}{|\vec{r}_j|} \right)^2 \tag{4.7}$$

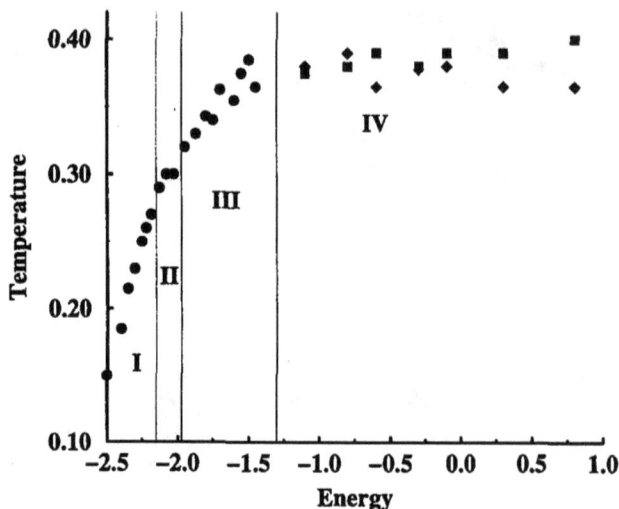

Fig. 4.13 Caloric curve for 100 Lennard-Jones disks that undergo fragmentation. In region *IV* the temperatures at τ_{ff} are represented by squares, and the asymptotic temperatures (of fragments with mass > 15) with diamonds.

where N_i is the total number of particles in region i in all events. In short we analyze the velocity fluctuations around the mean local collective velocity.

Caloric Curve After calculating the temperature of the concentric regions, the caloric curve can be constructed. Figure 4.13 shows the "extended caloric curve" (full squares), which encompasses the solid-like phase (region I), the liquid-like phase (region III), the associated phase transition (region II), as well as the high-energy process of evaporation and multi-fragmentation (region IV). The low energy parts of the caloric curves (regions I, II and III) were obtained using standard MD simulations.

Region IV shows the local temperature averaged over the three innermost regions from fragmentation computer experiments. They are also averaged over a time $t = 20t_0$ centered at the break-up time. This represents the break-up temperature of the system. It can be seen from figure 4.13 that the local temperature at break-up is quite independent of the total energy of the fragmenting system.

Connecting with Experiments. The last step in this analysis is the connection of these simulations with experiments, *i.e.* with asymptotic quantities. Since the asymptotic cluster excitation energy is accessible experimentally, at least in heavy-ion reactions and small atomic cluster experiments, it is natural to try to develop a connection to the hot and dense phase through this observable. Figure 4.13 shows the average internal temperature of clusters of mass > 15 at asymptotic times (full diamonds). The caloric curve can then be constructed directly from the temperature of the experimentally observable asymptotic fragments.

The fragmentation process can, accordingly, be divided in three stages. From $t = 0$ to τ_{ff} the radial flux and density fluctuations develop. These correspond to the cluster partition according to the $ECFM$ model. By the end of this stage the asymptotic fragments are already formed, although most of the mass of the system is still interacting and forming a big MST cluster. At τ_{ff} the value of the local temperature in the central region of the drop is equal to the limit temperature of its constituent clusters. The second stage of the process, goes from τ_{ff} to τ_{fe}, during which the fragments already formed separate in configuration space. The third and final stage, $t > \tau_{fe}$, corresponds to the free expansion stage. This scenario appears to be in agreement with experimental observations [Hauger *et al.* (1996)].

4.6 Conclusions

Phase transformations in nuclei are unique for the system under consideration is very small. Since the theory developed in the previous chapters is for large systems, modern technology to study critical phenomena in infinite systems was adapted for nuclear sizes in this chapter.

Critical exponents, order parameters, universality classes and *finite size scaling theory* were introduced to provide fingerprints of phase transitions in small systems. These novel concepts were tested using the *nuclear percolation model*, and the comparison of these results to infinite lattices, physical systems, and experimental nuclear data strongly suggests the possible existence of critical phenomena modulated by the finite size of the lattice.

To investigate dynamical aspects of fragmentation, *molecular dynamics* was used to study the disassembly of excited 2D Lennard-Jones small drops. The machinery to identify fragments, fragmentation time, and to determine the temperature had to be introduced. With this, the fragmentation process was found to be composed of three stages: an initial one where the radial flux and density fluctuations develop, an intermediate one during which the fragments separate, and a final one corresponding to a free expansion stage. Apparently the fragments can form early and can carry information about the dense and hot stage of the fragmentation process to the detectors.

The possibility of phase transformations in small systems was also verified by the caloric curve obtained from these MD simulations. The curve was found to have a "rise-plateau" shape with segments corresponding to a solid-like phase, a liquid-like phase, a phase transition, as well as an evaporation and multi-fragmentation segment.

Finally, it is interesting to see that most characteristics of infinite phase transformations survive in finite systems. Phases, order parameters, critical exponents and other physical objects from the world of *infinitia* continue to exist in the small systems limit. An important new ingredient found in disassembling finite systems is the radial collective motion, an extra degree of freedom which helps to deform the infinite physical quantities. As it will be seen in appendix A, this radial expansion has a strong impact on the non-linear aspects of fragmentation. All that remains now is to try to identify critical phenomena in nuclear experiments. This, up to a point, will be done in the following chapter.

Chapter 5

Heavy Ion Reactions

5.1 Background

This book would be nothing more than an academic exercise if there were not experimental data to corroborate the existence of phases in nuclear matter. The fact that nuclear matter behaves as a compressible liquid and can flow hydrodynamically has been known ever since the pioneering work of [Baumgardt *et al.* (1975)] and [Jakobsson *et al.* (1975)] with experiments involving medium and heavy ions colliding at kinetic energies of hundreds of MeV of energy per nucleon. But it was later that analysis of fragment mass distributions in proton- and heavy-ion induced reactions [Finn *et al.* (1982); Panagiotou *et al.* (1984)] found the first pieces of evidence for a nuclear liquid-gas phase transition.

This chapter presents a "minimalist" compilation of references to experimental facts that attest to the possibility of phase changes in nuclei. Supplementing this, a section on computational methods describes current phenomenological models from a very generic point of view. These methods and have been developed using a careful mix of theoretical and experimental ingredients, and are able to describe many experimental observations.

5.2 The General Scenario

As explained in section 3.1, a heavy-ion collision (HIC) could, in principle, form a compressed and heated system that might expand after the initial collision and enter into the coexistence or spinodals region breaking into droplets (fragments) and gaseous particles (light nuclei). Several considerations are worth reviewing.

- **Length Scale.** Taking the nuclear radius as $r = r_0 N^{1/3}$, a medium to heavy ion would have a diameter of $5 \sim 7 \; fm$. Since the saturation density is constant $\approx 0.15 \; fm^{-3}$ throughout the periodic table, the inter-nucleon distance is $\approx 2.3 \; fm$. [The range of the nuclear interaction is of the same magnitude.] Taking the nucleon's hard core radius as $\approx 0.5 \; fm$ [Walecka (1995)], the "packed spheres" density is $\approx 2 \; fm^{-3}$; this would be the maximum compression that will keep the nucleons not overlapping with one another (higher densities would necessitate the use of the quark degrees of freedom). [See problem 5.1.]

- **Energy Scale.** Since the binding energy of nuclei is $\approx 8 \; MeV/N$ (*i.e.* per particle), the energy needed for the partial disassembly of two colliding medium-size nuclei, say of $N \approx 100$, is on the order of one or two thousands of MeV. As this energy comes from the kinetic energy of the projectile, a beam energy of the order of a several tens of MeVs is needed.

- **Time Scale.** The time scale for the collision and possible thermalization of the participating nuclei is set by the kinetic energy of the projectile and the energy-transfer speed in nuclear matter. The energy-transfer time can be estimated using the high density value of the speed of sound, $c_s \approx 0.2 \; c$ (*cf.* figure 2.5), taking the distance traveled as twice the nuclear radius, the resulting time is $t \approx 2r_0 N^{1/3}/c_s \approx 55 \; fm/c$ or $18 \times 10^{-23} \; s$ for a $N = 100$ nuclei. The duration of the collision can be estimated as the time needed for the projectile to traverse the target. A $20 \; MeV$ particle would move at a speed of $v \approx 0.2 \; c$ making the collision time equal to the previous energy propagation time. [See problem 5.2.]

- **Caveats**

  Prompt Emission. The detection of a possible phase transition must rely on the analysis of the final particles produced in the reaction. But since particles can be emitted at any time of the reaction, not all of them can be taken as a possible signature for a change phase. For instance, in an initial contact in HIC, single nucleons can be knocked out directly by the projectile. These nucleons do not participate in the energy sharing and are not emitted as a phase-change byproduct.

Impact Parameter. Non-central collisions will reduce the number of participants in the thermal soup. The non-participant nucleons, *i.e.* the "spectators", will absorb varying amounts of energy which can be used to evaporate particles blurring possible phase transition signatures.

Particle Production. At the above mentioned energies, it is possible to produce pions and other particles. This reduces the energy going into the heating of the system inhibiting, perhaps, the breakup of nuclei.

Late Emission. After the disassembly of the hot and dense nuclear blob into droplets, excited fragments might shake off a few nucleons to "cool" down, again, contaminating the signature of those particles emitted during the phase change.

Quantum Effects. After bringing the temperature down by particle emission, the surviving nuclei will move toward isotopic stability by α and β decay. This process is certainly not directly related to the conditions that existed during the phase transition and will modify the final mass and isotope distribution.

Nuclear Temperature. Since the temperature of the colliding nuclei increases and decreases during the reaction, different temperature measurement techniques yield different results. For instance, the slopes of the kinetic energy spectra, can be used to estimate the temperature achieved by an evaporating hot source; this, presumably, would correspond to an early stage of the reaction. [See section 5.3.] Isotope yield ratios, on the other hand, can also reflect the temperature of the system when chemical equilibrium is achieved. But due to the longer time needed in achieving isotope balance, the resulting temperatures would be from a later stage. Indeed, the temperatures from these methods vary distinctively, see [Durand (1998)].

With this in mind, one can then adventure to apply the theory thus developed in hopes of seeing glimpses of a liquid-gas nuclear phase change. There are several pieces of experimental data that could be taken as evidence of a possible phase transition in heavy ion reactions. In particular, the kinetic energy and the fragment mass distribution of the emitted particles, and the caloric curve could provide possible evidence for thermalization, nucleation, and a first-order phase transition.

Problem 5.1   Lenght Scale

Using simple geometrical arguments, estimate the inter-nucleon distance existing at saturation density, and the "packed spheres" density assuming a nucleon hard core radius of $\approx 0.5 \ fm$.

Problem 5.2   Time Scale

Estimate the energy-transfer time in a $U + U$ collision assuming it reached $T \approx T_c$ at $\rho \approx 2\rho/3$, (cf. figure 2.5).

5.3   Kinetic Energy Spectra

An *a priori* requirement for the application of thermodynamics is the existence of thermal equilibrium in the nucleus during the reaction. If a nucleus thermalizes at some temperature, as explained in section 1.6, it can evaporate particles; this emission of particles can be detected experimentally and be used to confirm thermal equilibrium.

For sufficiently high temperatures, the kinetic energy distribution of the emitted particles (*cf.* section 1.6) goes from a Fermi-Dirac distribution to a Maxwell distribution. This has been observed experimentally, for instance, in the proton-induced fragmentation of Xenon [Porile *et al.* (1989)]. Figure 5.1 shows the energy spectra of *Be* (emitted at 48.5° off the proton beam) and the Maxwellian fit. Similar curves were obtained for a number of other isotopes.

A number of experimental considerations are worth being mentioned. Data collected includes collisions from different impact parameters, a factor not too important for proton beams. Low energy *Be* nuclei are not detected due to detector thresholds. Finally, since the particles are being emitted from a hot source which could be in motion, the kinetic energy distribution of the particles can be biased by this fact. To compensate for this, a "moving source" fit was used instead of a regular Maxwell fit.

The moving source fit is a convolution of a Maxwell distribution of a source with an excitation energy ϵ, and a temperature T_1, with a second Maxwellian that describes the distribution of the remaining energy, *i.e.* the center-of-mass energy, E_{cm}, minus the energy spent in the Coulomb repulsion, B, minus the excitation energy, and with a corresponding temperature parameter T_2:

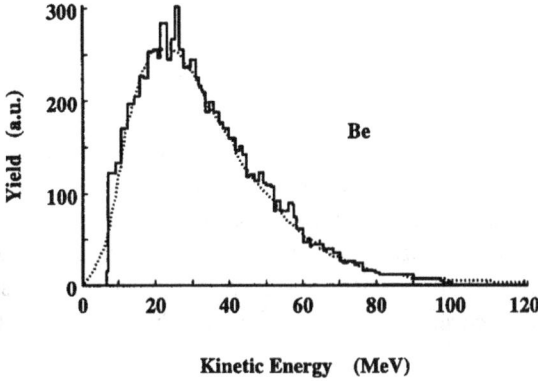

Fig. 5.1 Kinetic energy distribution of Be emitted in the reaction $p + Xe$ at proton energies between 10.1 and 11.4 GeV. Dashed line is a "moving-source" fit.

$$\frac{d^2\sigma}{d\Omega dE_{lab}} = N\sqrt{\frac{E_{lab}}{E_{cm}}} \int_0^{E_{min}} P(B)dB$$

$$\times \int_0^{E_{cm}-B} \sqrt{\epsilon(E_{cm}-B)} e^{-[\nu\epsilon/T_1]} e^{-[\nu(E_{cm}-B-\epsilon)/T_2]} d\epsilon .$$

Here N is a normalization factor, $\nu = (1 - A_f/A_r)^{-1}$ is a recoil-correction factor involving the mass numbers of the fragment and remnant, and $P(B)$ is the distribution of Coulomb repulsion energies. Given that the energy B is distributed, energy conservation demands that the first integral goes up to the smaller of E_{cm} or B_{max}.

To evaluate the integrals, a recipe to calculate B and its distribution is needed. This is necessarily model-dependent, and it involves calculating the interaction energy between the emitted fragment and the remnant at some distance apart, with this distance given by some distribution. Using one such model to fit the energy spectra of a number of fragment isotopes, [Porile et al. (1989)] obtained values from 12.0 to 15.6 MeV for the emitting temperature T_1, and from 2.2 to 4.4 MeV for T_2. Even though the resulting temperatures are a little on the high side for the model of nuclear matter used in the previous chapters, the obtained temperatures tell us, nevertheless, that we are in the correct "ball-park". For more recent data see [Borderie *et al.* (1999)].

5.4   Mass Yield

As explained in section 3.2.4, the mass distribution of the droplets can be related to the process of nucleation. Depending on whether the phase transition takes place on the supersaturated region, coexistence region, or at the critical point, the mass yield will have a characteristic dependence on the droplet size (*cf.* equations (3.8)-(3.10)). To use this as a signature of the phase change, a careful analysis of the experimental mass distribution is necessary.

Figure 5.2 shows three of the first observations of characteristic mass yields. The top curve is for proton-induced fragmentation of Krypton targets at beam energies of 80 to 350 GeV [Finn *et al.* (1982)], the middle and bottom ones are for Neon on Gold at 250 and 2100 MeV/N [Warwick *et al.* (1983)]. As seen in the figures, these yields can be well represented by a power-law fit, and were, initially, interpreted as indications of a critical phenomenon.

In principle, a pure power law should exist only in processes taking place at the critical point, and with a unique exponent τ independent of the system size, etc. (*cf.* equation (3.10)). The fact that figure 5.2 yields not a single τ, but a set ranging from 1.5 to 3.1, underlines the fact that not all reactions correspond to disassemblies at the critical point. This experimental data can be used, nevertheless, to obtain an indication of critical phenomena.

Using a power-law fit with data presumably produced around and at the critical point, a fitting parameter, τ, can be obtained. In fitting non-pure power law distributions, this "apparent exponent" effectively mimics the exponential factors of equations (3.8) and (3.9), and takes different values in the supersaturated and coexistence regions, or at the critical point. Since the exponential factors of these equations are smaller than 1 away from the critical point, the yield will be modulated by a reducing mass-dependent factor resulting in a steeper mass distribution and a larger τ. At the critical point, where the exponential functions ≈ 1, the yield should be less steep, and the apparent exponent should reach its smallest value. This procedure, then, should indeed allow us to obtain the real critical exponent and an

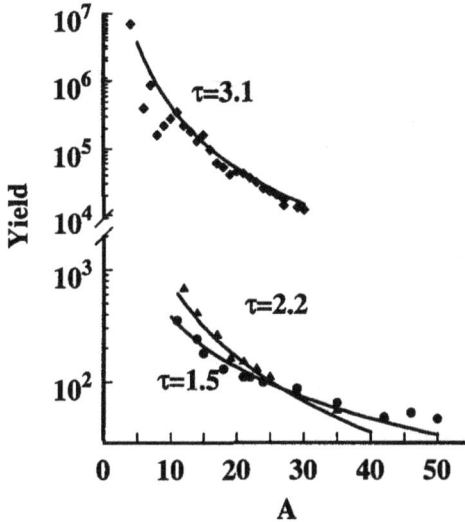

Fig. 5.2   Mass yield emitted in proton on Krypton at $80 - 350 \, GeV$ (top) and Neon on Gold at 250 and 2100 MeV/N.

estimation of the critical temperature from experimental data.

Figure 5.3 shows a compilation of apparent exponents obtained from a number of experiments. The solid circles show the values obtained from fitting the reactions $p+Ag$ (0.21−4.9 GeV), $p+U$ (4.9, 5.5 GeV), $p+Xe$, $p+Kr$ (80, 350 GeV), and $C+Ag$, $C+Au$ (0.18, 0.36 GeV) [Panagiotou et al. (1984)]. The empty circles show the values of τ obtained from the inverse-kinematics reactions $Au+C$, $Au+Al$, and $Au+Cu$ at 600 MeV/N [Ogilvie et al. (1991)]. The temperature dependence of the apparent exponents clearly shows a minimum of $\tau \approx 2$ at $T \approx 12 \, MeV$, again, in "ball park" agreement with theoretical expectations.

[For comparison purposes, the original data of [Ogilvie et al. (1991)], which was given in terms of deposited energy (E_{dep}), was naively transformed into "temperature" using $T = 2\kappa E_{dep}/3$, with κ adjusted to have

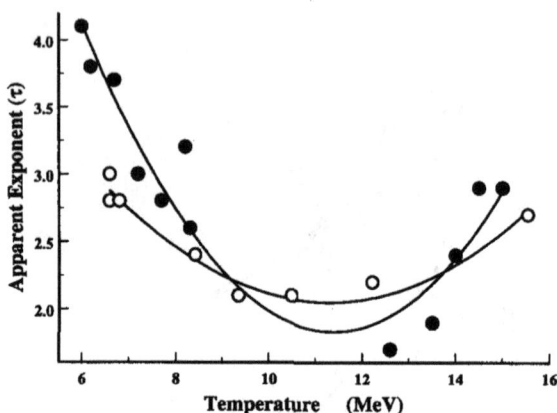

Fig. 5.3 Values of the apparent exponents τ obtained from several experiments.

the minimum of the data coincide with the compilation of [Panagiotou *et al.* (1984)].]

The temperature-variation of τ can be reasoned in terms of physics. At low energies, the excited compound nuclei merely has energy to evaporate a few light particles, this produces a steep mass distribution with some value of τ. As the energy increases, the heavy residues break and shrink populating the mid-charge section, this reduces the corresponding exponent. Finally, at higher energies, the intermediate mass fragments cool-off by evaporation producing more light particles and reducing the number of medium size fragments, this makes the charge distribution be steep again with a larger τ. This variation of the mass spectra has been nicknamed as the "rise and fall" of multi-fragmentation [Ogilvie *et al.* (1991)].

Although the charge distribution resulting from nuclear fragmentation can be fitted by a power law, and the τ exponents appear to have the correct T-variation, it is not guaranteed that the breakup process is a liquid-gas phase transition. On one hand, experiments keep on producing different values of the critical exponent (*eg.* $\tau \approx 2.2$ [Mastinu *et al.* (1996); D'agostino *et al.* (1995); Gilkes *et al.* (1994)] and other values [Trockel *et al.* (1989)]), on the other hand, researchers have argued that a finite expanding system should have modified decay laws [Bravina and Zabrodin (1996);

Li *et al.* (1994)]. As much as we would like to relate the charge distribution and its τ variation to critical phenomena, the question of whether a power law mass distribution identifies a nucleation process is still an open problem.

5.5 Caloric Curve

Another possible experimental signature of a phase transition might appear in the caloric curve of nuclear matter. As explained in section 1.5.1, energy added during a phase transition is used to break inter-particle bonds and produces, not an increase of the temperature, but a plateau (*c.f.* figure 4.10). Obtaining the caloric curve from a disassembling nucleus, however, is a demanding task as it requires the simultaneous extraction of the temperature achieved in the reaction as well as the excitation energy deposited in the nuclear system.

As mentioned in the list of caveats in section 5.2, the temperature achieved in a nuclear reaction evolves in time. To extract a phase transition signature from the caloric curve, the temperature used would have to be that of the nuclear system while the presumed transition takes place. Experimentally, however, this is hard to do as it is only the final products of the reaction that are detected. Pochodzalla and colleagues [Pochodzalla *et al.* (1995)] among others [Fabris *et el.* (1987); Kwiatkowski *et el.* (1998); Samaddar *et al.* (1997)] have used experimental data to construct such a caloric curve.

Figure 5.4 shows the so-called isotope temperature as a function of the excitation energy per nucleon obtained from the reactions $Au + Au$ at 600 MeV/N, $C, O + Ag, Au$ at 30 and 84 MeV/N, and $Ne + Ta$ at 8.1 MeV/N [Pochodzalla *et al.* (1995)]. The plateau signaling a first order phase transition is clearly visible at $T \approx 5\ MeV$ for $3 \leq E^* \leq 9\ MeV$. Also shown are the expected curves for a Fermi system, $E^* = aT^2$ (with the level density parameter adjusted to produce a best fit at $a = 1/12\ MeV^{-1}$), and the shifted classical expression $E^* = E_o + 3T/2$, with $E_o = 1.5\ MeV$.

In constructing this caloric curve, several assumptions were made. The imperfect isotope identification, for instance, introduced an uncertainty in the excitation energy which required a correction based on a random sampling. Also in evaluating the fragment's kinetic energies, an isotropic decay was assumed and a Coulomb correction was introduced.

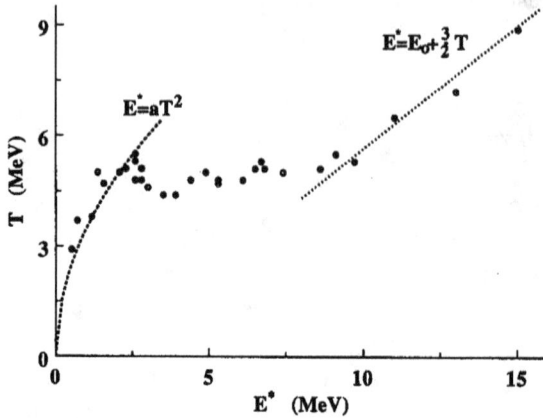

Fig. 5.4 Experimental determination of the nuclear caloric curve

As mentioned before, the temperature was obtained, not using the energy spectra method of section 5.3, but the isotope ratio technique. This method assumes the existence of both thermal and chemical equilibrium during the reaction, and uses the isotope ratios to estimate a temperature [Albergo *et al.* (1985); Tsang (1997)]. As the isotope ratios might reach their asymptotic values in the final stages of the reaction, it could be that the temperature they reflect are lower than those existing during the possible phase transition. This could explain why the values of T observed in figure 5.4 are lower than those extracted from kinetic energy spectra as in figure 5.1.

[The fact that more accurate thermometers are needed is exemplified by analyses of the same data that use only the Li^5 isotopes and produce the "rise plateau" pattern without the "classical" final rise [Serfling *et al.* (1998)]. This is in agreement with the fragmentation of Lennard Jones drops shown in figure 4.13 of § 4.5.3.]

As exemplified by the curves shown, the $T - E^*$ relationship appears to follow the predictions for a Fermi liquid at low energies, and that of a classical gas at higher energies. In previous works [Fabris *et el.* (1987)], similar Fermi-like behavior has been observed with varying values of the level density parameter. Pochodzalla and colleagues argue that the shift of

$E_o = 1.5 \; MeV$ might be related to the fact that the system loses thermodynamic contact at some finite "freeze-out" density during the final expansion.

With all these reservations, the apparent transition from a Fermi-liquid behavior at low energies to a classical gas at higher energies, mediated by a constant-T plateau seems to indicate that a liquid-gas phase transition is indeed possible in a nuclear reaction.

Problem 5.3   Nuclear Density at Disassembly
Identify the curve $E^* = aT^2$ of figure 5.4 with $E_T(T)/N$ of section 1.3.1, and use the fitted value of $a = 1/12 \; MeV^{-1}$ and ϵ_F from problem 1.2 to obtain the density of the disassembling system at energies in the range $0 \leq E^* \leq 3 \; MeV$ where the Fermi curve appears to be valid.

5.6   Computational Methods

In this book, nuclear matter was studied first as a non-interacting infinite nucleon gas, then as an infinite interacting system with a simple nuclear equation of state. This paved the road to study possible phase transitions and to generalize to critical phenomena in finite systems. But it was not until the real complexity of heavy ion reactions was examined in more detail. The theory of the previous chapters, as extensive as it is, does not provide a complete framework to study the dynamical problem of nuclear multi-fragmentation.

As two heavy nuclei approach each other in a collision, they form a highly-correlated strongly-interacting finite quantum system. As they collide and the beam energy is distributed, the participant nucleons reach higher energy levels where the Pauli principle is less restrictive on the energy transfer. At this point the nuclear Fermi-gas begins to resemble a more classical one, and thus one is allowed to speak of temperatures and phase transitions in the classical sense. After a presumed phase transition takes place, the system expands and disassembles into clusters (liquid droplets and gaseous particles), which continue de-exciting by further fission and photon and particle emission.

This transition from a quantum regime to a more classical one, and back again to the quantum world, makes heavy-ion reactions practically impossible to track analytically. Apart from being a many-body system (for which no exact solutions exist), the reaction dynamics are highly out-of-equilibrium and required the use of quantum kinetic theory which does not exist yet. Fortunately, since the collision can be divided into, say, three stages, collision/compression, expansion/breakup, and cooling, computational models with some predictive power have been devised.

These methods can be broadly divided into dynamical and statistical models, with some hybrid approaches linking these two extremes. As there exists an extensive literature on most of these models, the following sections present some of these approaches in the briefest form possible.

5.6.1 *Dynamical Models*

In a perfect world, a theory describing the quantum behavior of every nucleon and including all particle-particle correlations would be able to describe many-particle processes such as changes of phase and cluster formation. However in our imperfect world this is not possible and approximations are invariably introduced.

Boltzmann-Uehling-Uhlenbeck Increasing in complexity from a single-particle description (such as that of a chapter 1) mean fields can be introduced to modulate the overall behavior of nucleons. In the Boltzmann-Uehling-Uhlenbeck (*BUU*) model [Bertsch and das Gupta (1988)], the heavy ion collision is described by the temporal evolution of the one-body density within a mean field. This mean field is usually taken as that produced by the motion of the nucleons averaged over an ensemble of configurations.

Other related approaches are the Vlasov-Uehling-Uhlenbeck [?] and the Boltzmann-Langevin [Ayik *et al.* (1996)] methods. As all these techniques use only a mean field ignoring higher-order correlations, these approaches, along with *BUU*, are not able to explore cluster formation. Refined models introduce mean field fluctuations to artificially induce clustering [Randrup and Ayik (1994)], or are used in hybrid models to add an *ad hoc* fragmentation [Botvina *et al.* (1990); Wang *et al.* (1999)].

Semiclassical Molecular Dynamics As explained in section 4.5, molecular dynamics calculations solve the classical equations of motion on the colliding nucleons using mostly r-dependent two-body potentials. Stemming from the original work of [Wilets (1978)], several models have incorporated different features [Boal and Glosgli (1988); López and Lübeck (1989); Dorso *et al.* (1993)] . In particular in a series of works [Dorso and Randrup (1987)], a model, dubbed QCNM (Quasi Classical Nuclear Model), was developed in which the Pauli exclusion principle is simulated by using an interaction potential in phase space (i.e. a Gaussian repulsive potential that depends on the relative momentum and the relative coordinate). More recently a simpler approach that incorporates this ideas has been developed which we refer as "Semiclassical Molecular Dynamics" (SMD) of [Pandharipande *et al.* (1990)].

Using a potential that appropriately reproduces the nuclear binding energies and radii and the energy-dependent nucleon cross section, SMD gives a complete, albeit classical, description of every nucleon. Incorporating all particle correlations, the model can correctly reproduce collective nuclear motion, ranging from hydrodynamics to critical phenomena. Although the model has been used to study the isentropic expansion of a compressed nucleus [Pandharipande *et al.* (1990)], its predictions in collisions are still being uncovered [Barrañón *et al.* (1999); Chernomoretz (2000)].

Quantum Molecular Dynamics The "Quantum Molecular Dynamics" model (QMD) [Aichelin (1991); Peilert *et al.* (1992)] attempts to manipulate a many-body wave function in hopes of introducing the correlations needed for cluster formation. Taking the wave function as a product of coherent states and using momentum-dependent interactions with Pauli blocking, the model evolves a set of test particles through the temporal evolution of the reaction.

Much like BUU, This model yields a detailed picture of the hot and dense stage of the reaction providing valuable insight on the time and energy scales. Unfortunately, besides of being computationally intensive, QMD fails to reproduce large fragment multiplicities due, most likely, to its *ad hoc* introduction of the Pauli principle as well as the lack of fluctuations in the model. In an attempt to improve the treatment of Pauli principle the Gaussian repulsion potential above mentioned has been introduced in this model[Peilert *et al.* (1992)]. In an attempt to squeeze information out of the pre-fragment's density and excitation energy provided by QMD, it has been coupled to statistical models to produce final mass and energy yields [Tsang *et al.* (1993)].

5.6.2 *Statistical Models*

The statistical fragmentation models, such as that of [Koonin and Randrup (1987)], SMM [Bondorf *et al.* (1995); Botvina *et al.* (1990)], or $MMMC$ [Gross (1997)] assume the formation of a hot and dense blob of nuclear matter that disassembles simultaneously into *non-interacting* fragments and light particles. Using different approximations, these models use "freeze-out" volumes as parameters, and invoke statistical arguments, which implicitly assume that the system is at equilibrium, to calculate the most probable mass and energy partitions. Without making explicit reference to the breakup dynamics, these models can assign thermal and Coulomb motion to the produced fragments to mimic a post-breakup dynamics. With varying degrees of success, these models reproduce inclusive experimental data such as mass and kinetic energy spectra, and predict interesting features of the equation of state, such as the caloric curve.

The omission of inter-fragment interactions was corrected in the "Transition State Treatment of multi-fragmentation" (TST) of [López and Randrup (1994)]. In generalizing the transition state model of binary fission to muti-fragment fission, a fragment-fragment potential was introduced with the additional advantage of removing the freeze-out volume parameter and introducing the post-breakup dynamics in a natural way.

Using a different approach, the "Expanding Emitting Source" (EES) model [Friedman (1990)] partially corrected the lack of built-in dynamics of the statistical models. This model assumes an oscillating hot source which, at low energies, emits light particles, and, at higher energies, generates large-amplitude oscillations that exceed the attractive nuclear forces disassembling the source. In spite of introducing a radial expansion, the kinetic spectra of the fragments appears to be not yet fully explained by this or the previous statistical models [Rivet *et al.* (1999)]. For a discussion on other models see [Botet and Ploszajczak (1994)].

5.6.3 *Sequential Models*

A third category of approaches is the one based on a sequential production of fragments. Since at low energies this mode of decay is known to be dominant, it is justified to expect its survival at higher energies. This type of decay, unfortunately, obscures the thermal history of the reaction, and prevents us from exploring the hotter stage of the collision through an analysis of the final products [López and Randrup (1989)]. A model mostly used for light particle emission is the code *Lilita* [Gómez del Campo (1979)], while one that includes emission of all size particles is *Gemini* [Charity *et al.* (1988)]. A computational code that allows a comparison of sequential with simultaneous emission is SOS [López and Randrup (1992)].

5.7 Conclusions

This chapter represents a minimal effort to corroborate the existence of phases in nuclear matter using experimental data. To examine such possibility, a general scenario was adopted in which a heavy-ion collision forms a compressed and heated system that expands into the coexistence or spinodals regions disassembling into fragments and light nuclei. Considerations regarding length, energy and time scales were reviewed as well as some caveats related to prompt and late particle emission, the role of impact parameter, quantum effects, and absence of reliable nuclear thermometers.

With this in mind, particles emitted in an experimental reaction could be attributed to a possible evaporation from a thermalized source. To verify that indeed this hypothesis is true, the distribution of the kinetic energy of the detected particles was compared successfully to a Maxwell distribution. This can be interpreted as a signal for thermalization at temperatures in the range of $T \approx 12 - 15\,MeVs$.

Next, if a thermalized hot source is being produced, processes such as nucleation could also be at work. Remarkably, experimental mass yields from proton-induced fragmentation of Krypton can be well represented by a power-law fit, as predicted by nucleation theory. And since the extracted power-law exponent was found to show a temperature variation, this can be indicative of nucleation occurring in different regions of the T-n plane.

Finally, if nucleation is taking place, the formation of phases should be reflected in the caloric curve. A careful extraction of the excitation energy used in an experimental reaction and the temperature achieved, allows the calculation of the caloric curve of nuclear matter. Experimental results seem to indicate that for a large energy range, the energy added does not increase the temperature forming the plateau expected in phase changes.

In summary, experimental data appears to support the existence of a liquid-gas phase transition in nuclear collisions. Thermalization seems to be established during the reaction, nucleation appears to be responsible for particle production, and preliminary caloric curves seem to indicate that liquid-gas phase transitions can indeed occur in experimental heavy-ion collisions. A rather surprising result that contrasts the macroscopic aspects of phase transitions with the microscopic size of a nucleus, literally *"ad augusta per angusta"*.

Appendix A

Non-Linear Aspects of Fragmentation

A.1 Introduction

As seen in chapters 4 and 5, the process of fragmentation of small systems is characterized by a caloric curve that displays a plateau. This feature can be attributed to the fact in that region the energy is used to break the system instead of heating it up. Part of the energy goes into producing small fragments, while the remaining energy is used to build up the collective radial flux. This process was studied microscopically in chapter 4 following the highly non-linear evolution using a Lennard-Jones interaction potential. In this appendix, the chaotic properties of such a phenomena will be analyzed making use of the Lyapunov exponents (LE).

A characteristic property of chaotic motion is the high sensibility to the initial conditions. Chaotic close neighboring trajectories diverge exponentially in time, but regular trajectories, on the other hand, diverge only linearly. The mathematical objects that properly quantify the rate of exponential divergence are the Lyapunov exponents. For mathematical proofs and formal definitions see [Lichtemberg and Lieberman (1992); Beck and Schlogl (1997)].

A.2 Lyapunov Exponents

Consider a system governed by the following set of differential equations

$$\frac{dx_i}{dt} = F_i(x_1, \cdots, x_n)$$

with $i = 1, \cdots, n$. Linearizing the equations of motion around a reference trajectory in n-dimensional phase-space, $\mathbf{x} = (x_1, \ldots, x_n)$, produces the following tangent map

$$\frac{d\delta x_i}{dt} = \sum_i^n \delta x_i \left(\frac{\partial F_i}{\partial x_i}\right)_{x=\bar{x}}$$

The Euclidean norm $d^2(\mathbf{x}(0)t) = \sum_i^n \delta x_i^2(\mathbf{x}(0)t)$ can be used to measure the divergence between a reference trajectory $\mathbf{x}(0)$ and an initially neighboring one $\mathbf{x}'(0) = \mathbf{x}(0) + \delta\mathbf{x}(0)$. The mean rate of divergence can then be quantified by

$$\lambda(\mathbf{x}(0), \delta\mathbf{x}(0)) = \lim_{t \to \infty} \lim_{d_0 \to 0} \frac{1}{t} \ln\left(\frac{d(\mathbf{x}(0), t)}{d(\mathbf{x}(0), 0)}\right) \qquad (A.1)$$

It can be shown that there exists an n-dimensional basis $\{\varepsilon_i\}$ of $\delta\mathbf{x}$ such that for any $\delta\mathbf{x}$, λ takes on one of the possibly distinct n values, the so-called Lyapunov characteristic exponents. The values of these coefficients are independent of the metric used for the phase-space and can be ordered by size: $\lambda_1 \geq \lambda_2 \geq \lambda_3 \geq \cdots \geq \lambda_n$.

Liouville's theorem imposes a constraint among the Lyapunov exponents. Consider a $6n$-dimensional phase-space sphere centered around the initial condition $\bar{x}(0)$. As this sphere evolves in time and deforms into an ellipsoid, its volume can be described as

$$V(t) = V(0) \exp\left(\sum_i \lambda_i t\right) .$$

But since this is a Hamiltonian system, according to Liouville's theorem, its volume in phase-space is to be conserved and, thus, the exponent should be zero. In this case, the Lyapunov exponents come in pairs, each composed of one positive exponent related to expansion in phase-space and one negative which, in turn, takes care of the shrinking.

The exponent that rapidly dominates the overall deformation process is called the "Maximum Lyapunov Exponent", or MLE. This exponent has been used by many researchers as a test for stochasticity [Ford (1975); Benettin *et al.* (1976); Shimada and Nagashima (1979)] and in the study of solid-like to liquid-like transitions in finite systems [Nayak *et al.* (1995)]. Since the calculation of the complete spectrum of Lyapunov exponents is a very demanding task, only the MLE will be treated here.

To apply these ideas to the fragmentation of excited drops, first an Euclidean norm in phase-space will be defined. Building an initial configuration of a system composed of A particles and with a given excitation energy, a point in phase-space can be defined by

$$(\mathbf{q}_0, \mathbf{p}_0) = (q_x^1, q_y^1, q_z^1, p_x^1, p_y^1, p_z^1, \cdots, q_x^A, q_y^A, q_z^A, p_x^A, p_y^A, p_z^A) \,.$$

A "son" can now be generated by slightly modifying this initial condition, *i.e.* taking one of the coordinates at random and changing it by a small amount, say $p_x^l \rightarrow p_x^l(1 + 10^{-6})$. A distance in phase-space, $d(t)$, can be defined as

$$d^2(t) = \sum_{i=1}^{N} [a(\mathbf{q}_0(t) - \mathbf{q}_s(t))^2 + b(\mathbf{p}_0(t)) - \mathbf{p}_s(t))^2]_i \,,$$

where q and p are the positions and momenta of N particles at time t, the indices 0 and s refer to the main and son trajectories (which initially differ by $d(0)$), and a and b are two arbitrary parameters which take care of dimensions. Since the LE are independent of the particular metric used in phase-space [Oseledec (1968)], a and b can be chosen as $a = 0$ and $b = 1/m^2$ (m being the mass of the particles) to use only distances in velocity-space.

When calculating numerically the time evolution of $d(t)$ by solving the classical equations of motion, on average, an exponential increase followed by a saturation in velocity-space is observed. Figure A.1 shows the result of one of these calculations for an expanding Lennard-Jones drop such as those used in section 4.5. The saturation effect is readily noticeable, however, the information contained in the slope of this curve is not as good as it can be as it averages a lot of different behaviors of the exploding system. It includes, for instance, information from the very early, dense, hot and highly collisional stage of the evolution, as well as from later stages in which the system has developed a collective radial motion.

This situation can be improved as follows. Dividing the total time in N "steps" of size Δt, equation (A.1) can be written as

$$\begin{aligned}
\lambda &= \lim_{N \to \infty} \lim_{d(0) \to 0} (\frac{1}{N\Delta t}) \ln \left[\frac{d(\Delta t)}{d(0)} \frac{d(2\Delta t)}{d(\Delta t)} \cdots \frac{d(t)}{d(t - \Delta t)} \right] \\
&= \lim_{N \to \infty} \lim_{d(0) \to 0} (\frac{1}{N}) \left\{ \sum_{i=1}^{N} \frac{1}{\Delta t} \ln \left[\frac{d(i\Delta t)}{d((i - 1)\Delta t)} \cdots \right] \right\} \,,
\end{aligned}$$

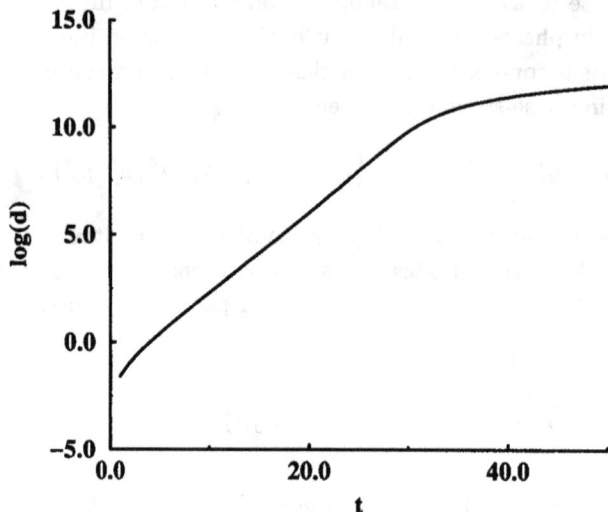

Fig. A.1   Time evolution of the distance between two neighboring trajectories of a frag-menting Lennard-Jones drop. After an initial steep rise, the curve "saturates" reflecting the finiteness of the phase-space available for an exploding drop.

which can be interpreted as an average over local (in time) Lyapunov expo-nents. This suggests that to obtain information relevant to the fragmenta-tion problem, it is necessary to look at the larger contributions to the LEs at the proper intervals. For a non-stationary system, as in the case of frag-mentation, we have to look at the "Maximum Local Lyapunov Exponent" ($MLLE$), which is closely related to the MLE.

A.3   Maximum Local Lyapunov Exponent

A practical algorithm to calculate MLE was devised by [Benettin *et al.* (1976)], and consists of calculating the quantity $d(\Delta t)$ by integrating the equations of motion for a small fixed interval Δt, and taking the initial norm $d(0) = 1$. At time Δt the distance is renormalized again to one, the process is repeated N times, and the quantity $\ln[d_1/d_0]$ is saved.

To implement this, a set of initial conditions (ensemble) consistent with the macroscopic state of the system is first built. In this case all conditions

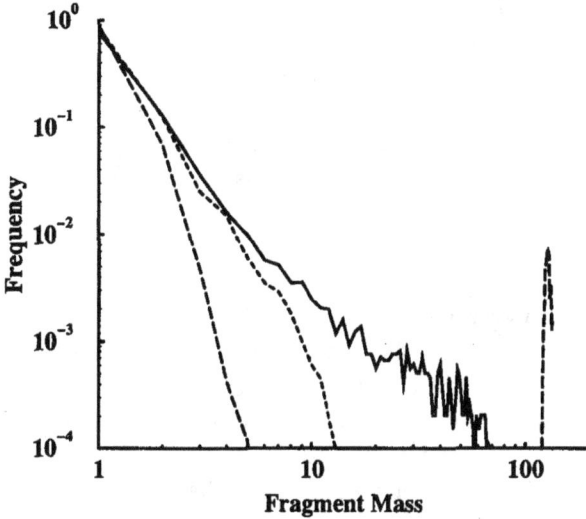

Fig. A.2 Mass spectra produced by the fragmenting 147-particle Lennard Jones drop. Full line denotes results for $\epsilon = 0.5\,\epsilon_0$, and the resulting spectrum is power-law like. The short-dashed line corresponds to $\epsilon = -2\,\epsilon_0$ with a U-shaped spectrum. The long-dashed line is for $\epsilon = 3\,\epsilon_0$ with an exponentially decaying spectrum.

have the same excitation energy. Second, for each initial condition a set of sons is constructed as described before. Next, each set of original points and sons are evolved in M_i small time steps of duration τ each, leaving them at time $t_i = \sum_i M_i \tau$. This effectively replaces the former time interval by $\Delta t = M_i \tau$.

The maximum local Lyapunov exponent associated with the ith interval of size $M_i \tau$ and beginning at $t_i = \sum_i M_i \tau$ can then be defined as

$$\lambda_i = \lambda(t_i) = \frac{1}{M_i \tau} \sum_{j=i}^{i+M_i} ln \left| \frac{d_{j+1}}{d_j} \right| = \frac{1}{M_i \tau} ln \left| \frac{d_{i+M_i}}{d_i} \right| . \qquad (A.2)$$

This, which is valid for each main point, is then averaged over the ensemble of time-evolved initial conditions corresponding to the set of trajectories macroscopically equivalent (*i.e.* with the same energy) at $t = 0$. It should be emphasized that the MLE is defined as an average over an infinitely long trajectory, while the $MLLE$ is an average over an ensemble of macroscopically equivalent initial conditions over a finite time span.

The maximum local Lyapunov exponent can also be calculated with a slightly different method. It also consists in generating a set of initial conditions that evolve in phase-space, but after a period of time $\tau \gg \lambda^{-1}$, a new set of sons is randomly generated erasing all previous information. In this way, different regions of the phase-space visited by the system can be characterized with the $MLLE$ calculated as,

$$\lambda_i = \lambda(t_i) = \frac{1}{N_i \tau} \ln \left[d(t_i + M_i \tau)/d(t_i) \right] , \qquad (A.3)$$

where, once again, an average over macroscopically equivalent main points is performed. We have found that both ways of calculating the $MLLE$ (using equations (A.2) and (A.3)) give equivalent results.

These ideas can be now be tested using the same molecular dynamics methodology used in section 4.5 for the analysis of two-dimensional drops, but now for 3D drops. Figure A.2 shows the mass spectra resulting from numerical simulations of the disassembly of 3D Lennard-Jones drops composed of 147 particles for three different energies per particle, namely $\epsilon = -2.0\,\epsilon_0$, $+0.5\,\epsilon_0$, and $+3.0\,\epsilon_0$. Notice that 147 is a magic number, *i.e.* it is a closed shell arrangement of particles, and the ground state energy for this system is $E = -5.8\,\epsilon_0$.

For $E = -2.0\,\epsilon_0$ (figure A.2 dashed line), the asymptotic spectrum consist in a big fragment of about 130 particles plus some free particles (U-shaped mass spectra). The mass spectra of $E = +0.5\,\epsilon_0$ (full line) is power law like, and the one corresponding to $E = +3.0\,\epsilon_0$ (long-dashed line) is mainly composed by small clusters. The $MLLE$'s can then be calculated using equation (A.2) with $\tau = 0.01\,t_0$ and $N_i = 1000$, so we have one point every $10\,t_0$, except for the first $10\,t_0$ of the evolution for which we use $N_i = 100$ and then $N_i \tau = 1\,t_0$.

The behavior of the $MLLE$ as a function of the time can be seen in figure A.3 for the same three energies analyzed in figure A.2. It can be seen that in all cases the behavior is the same, the $MLLE$s decrease sharply in the first $20\,t_0$ of the evolution, reaching an almost stationary value after this time.

This behavior shows us that we can classify the time evolution of the system in two well differentiated stages, the early one, when the chaoticity of the drop is decreasing quickly, and the asymptotic one, characterized by the stability of the values of the $MLLE$. We can define a characteristic time, τ_{mle}, as the time at which the $MLLE$ reach their asymptotic values.

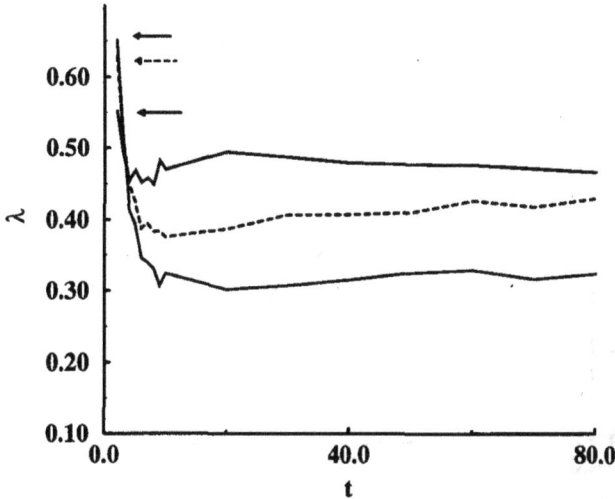

Fig. A.3 Lyapunov exponents as a function of time for three values of the energy of the system. Full lines stand for $\epsilon = -2.0\epsilon_0$, dotted line for $\epsilon = 0.5\,\epsilon_0$ and dashed line for $\epsilon = 3\,\epsilon_0$. The arrows are included to help identify the $t = 0$ values of the Lypunov exponents.

In this system $\tau_{mle} = 20\,t_0$.

It can also be seen that the energy dependence is different at very early times than at asymptotic times. To illustrate this, figure A.4 plots the $MLLE$ as function of the energy at three relevant times. At very early times, *i.e.* $t = 2\,t_0$, the $MLLE$ is an increasing function of the energy. This behavior can be understood because the drop has not developed the radial flux yet, it is still compressed and fragments are not formed yet. Due to the absence of collective radial motion all the excitation energy goes into what could be called "chaotic motion of the particles" and the system behaves as an infinite one. This can be denoted as "infinite system like" behavior.

On the other hand, for $t \geq 20\,t_0$ the system is fully in the asymptotic regime and the $MLLE$ displays a maximum. This asymptotic behavior depends mainly on the size and the temperature of the biggest fragment of the fragmented drop. Looking at the mass spectra in figure A.2, it can be seen that at the energy for which the maximum of the $MLLE$ occurs, the

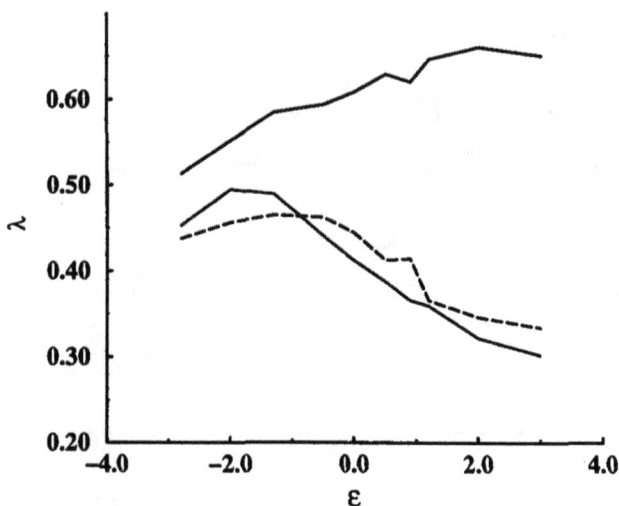

Fig. A.4 Lyapunov exponents as a function of the energy at three typical times, full line for $t = 2\,t0$, dotted line for $t = 20\,t0$, dashed line for $t = 150\,t0$.

asymptotic biggest fragment is as big as the whole system. In this case one can talk about "finite size like" behavior, the asymptotic $MLLE$ brings information about the dominant fragment in the asymptotic regime.

A.4 Chaos to Order Transition

To get some insight into the reasons why the $MLLE$'s behave the way they do, one could explore the strength of the collisional process undergone by the constituents particles. This can be accomplished calculating the mean momentum transfer between particles (MMT) as a function of the energy and time ($C(t)$). MMT is defined as

$$C(t) = \frac{1}{N_{conf}} \sum_{j=1}^{N_{conf}} \sum_{i=1}^{N} m.\,|\vec{v}(t+\delta) - \vec{v}(t)| \qquad (A.4)$$

where N_{conf} is the number of configurations analyzed for each energy, i.e., an average over the ensemble of initial conditions.

The time dependence of the MMT is shown in figure A.5 for the same

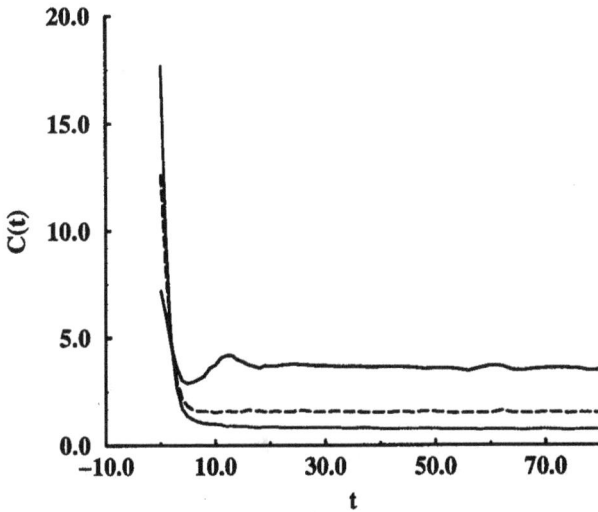

Fig. A.5   Mean momentum transfer as a function of time for the same energies as in the previous figures.

three excitation energies analyzed in figure A.3. The MMT behaves in a similar way as the $MLLE$, at early times the collisions increase with energy, but they decrease very fast as a function of time. $C(t)$ finally reaches its almost stable asymptotic values.

In this context, it can be seen that the $MLLE$ are strongly dependent on the interparticle collisions, which in turn depend on the collective radial motion. In brief, at very early times the drop is highly chaotic because all the excitation energies goes to inter-particle collisions (the radial flux is not formed yet); but when the system expands, the collisions between particles decrease and the system becomes less chaotic. This process arrives at a stable asymptotic regime when the radial collective mode is fully formed. It is then clear that the system begins in a highly chaotic state and "decays" to a more ordered one. In this way fragmentation can be viewed as a *"chaos to order transition."*

Appendix B

Computer Codes

B.1 Fermigas.for

This fortran program calculates the density, pressure, energy and entropy of a nuclear Fermi-gas at a given T and μ. It uses subroutine ferdr.

```
1   write(6,*)'input T and mu, T<0 => exit'
    read(5,*)t,xmu
    if(t.le.0.)goto 2
    eta= xmu/t
    call fermi(t,eta,d,s,p,e)
    goto 1
2   stop
    end
    subroutine fermi(t,eta,d,s,p,e)
    data factor/1.0768409e-3/
    fdd=ferdr(eta,1)       !fermi-dirac integral for the density
    fde=ferdr(eta,3)       !fermi-dirac integral for the pressure
    d=factor*t**1.5*fdd    !calculate the density from fermi integral
    if(d.eq.0.)d=1.e-9
    p=2./3.*factor*t**2.5*fde    !calculate the pressure
    if(fdd.eq.0.)fdd=1.e-9
    e=t*fde/fdd            !calculate the energy per nucleon
    s=p/d/t + e/t - eta    !entropy from fermi integral
    write(6,*)t,eta,d,s,p,e
    return
    end
```

B.2   Ferdr.for

This fortran subroutine evaluates Fermi-Dirac integrals of the form
$\int_0^\infty t^{k/2} dt/(1 + e^{t-x})$ with integer $k = -1, 1\,3$ and real x. For details
see *CERN* subroutine *C*323 at wwwinfo.cern.ch or [Cody and Thacher
(1967)].

```
FUNCTION FERDR(RX,K)
DOUBLE PRECISION DFERDR,DX,X,Y,H
DOUBLE PRECISION P1,P2,P3,P4,P5,P6,P7,P8,P9
DOUBLE PRECISION Q1,Q2,Q3,Q4,Q5,Q6,Q7,Q8,Q9
DOUBLE PRECISION C1,C2,C3,C4,C5
DOUBLE PRECISION ZERO,ONE,TWO,FOUR
CHARACTER NAME*6
DIMENSION P1(5),P2(5),P3(5),P4(5),P5(5),P6(5),P7(5),P8(5),P9(5)
DIMENSION Q1(5),Q2(5),Q3(5),Q4(5),Q5(5),Q6(5),Q7(5),Q8(5),Q9(5)
DATA C1 /1.77245 38509 05516 03D0/
DATA C2 /0.88622 69254 52758 01D0/
DATA C3 /1.32934 03881 79137 02D0/
DATA C4 /0.66666 66666 66666 67D0/
DATA C5 /0.40000 00000 00000 00D0/
DATA ZERO /0D0/, ONE /1D0/, TWO /2D0/, FOUR /4D0/
DATA P1
1/-1.25331 41288 20D+0, -1.72366 35577 01D+0, -6.55904 57292 58D-1,
2 -6.34228 31976 82D-2, -1.48838 31061 16D-5/
DATA Q1
1/+1.00000 00000 00D+0, +2.19178 09259 80D+0, +1.60581 29554 06D+0,
2 +4.44366 95274 81D-1, +3.62423 22881 12D-2/
DATA P2
1/-3.13328 53055 70D-1, -4.16187 38522 93D-1, -1.50220 84005 88D-1,
2 -1.33957 93751 73D-2, -1.51335 07001 38D-5/
DATA Q2
1/+1.00000 00000 00D+0, +1.87260 86759 02D+0, +1.14520 44465 78D+0,
2 +2.57022 55875 73D-1, +1.63990 25435 68D-2/
DATA P3
1/-2.34996 39854 06D-1, -2.92737 36375 47D-1, -9.88309 75887 38D-2,
2 -8.25138 63795 51D-3, -1.87438 41532 23D-5/
DATA Q3
1/+1.00000 00000 00D+0, +1.60859 71091 46D+0, +8.27528 95308 80D-1,
```

```
2 +1.52232 23828 50D-1, +7.69512 04750 64D-3/
      DATA P4
1/+1.07381 27694D+0, +5.60033 03660D+0, +3.68822 11270D+0,
2 +1.17433 92816D+0, +2.36419 35527D-1/
      DATA Q4
1/+1.00000 00000D+0, +4.60318 40667D+0, +4.30759 10674D-1,
2 +4.21511 32145D-1, +1.18326 01601D-2/
      DATA P5
1/+6.78176 62666 0D-1, +6.33124 01791 0D-1, +2.94479 65177 2D-1,
2 +8.01320 71141 9D-2, +1.33918 21294 0D-2/
      DATA Q5
1/+1.00000 00000 0D+0, +1.43740 40039 7D-1, +7.08662 14845 0D-2,
2 +2.34579 49473 5D-3, -1.29449 92883 5D-5/
      DATA P6
1/+1.15302 13402D+0, +1.05915 58972D+0, +4.68988 03095D-1,
2 +1.18829 08784D-1, +1.94387 55787D-2/
      DATA Q6
1/+1.00000 00000D+0, +3.73489 53841D-2, +2.32484 58137D-2,
2 -1.37667 70874D-3, +4.64663 92781D-5/
      DATA P7
1/-8.22255 9330D-1, -3.62036 9345D+1, -3.01538 5410D+3,
2 -7.04987 1579D+4, -5.69814 5924D+4/
      DATA Q7
1/+1.00000 0000D+0, +3.93568 9841D+1, +3.56875 6266D+3,
2 +4.18189 3625D+4, +3.38513 8907D+5/
      DATA P8
1/+8.22449 97626D-1, +2.00463 03393D+1, +1.82680 93446D+3,
2 +1.22265 30374D+4, +1.40407 50092D+5/
      DATA Q8
1/+1.00000 00000D+0, +2.34862 07659D+1, +2.20134 83743D+3,
1 +1.14426 73596D+4, +1.65847 15900D+5/
      DATA P9
1/+2.46740 02368 4D+0, +2.19167 58236 8D+2, +1.23829 37907 5D+4,
2 +2.20667 72496 8D+5, +8.49442 92003 4D+5/
      DATA Q9
1/+1.00000 00000 0D+0, +8.91125 14061 9D+1, +5.04575 66966 7D+3,
2 +9.09075 94630 4D+4, +3.89960 91564 1D+5/
      X=RX
```

```
      NAME='FERDR '
      FERDR=ZERO
    9 IF(K .EQ. -1) THEN
        IF(X .LE. ONE) THEN
         Y=EXP(X)
         H=Y*(C1+Y*
    1 (P1(1)+Y*(P1(2)+Y*(P1(3)+Y*(P1(4)+Y*P1(5)))))/
    2 (Q1(1)+Y*(Q1(2)+Y*(Q1(3)+Y*(Q1(4)+Y*Q1(5))))))
        ELSE IF(X .LE. FOUR) THEN
         H=(P4(1)+X*(P4(2)+X*(P4(3)+X*(P4(4)+X*P4(5)))))/
    1 (Q4(1)+X*(Q4(2)+X*(Q4(3)+X*(Q4(4)+X*Q4(5)))))
        ELSE
         Y=ONE/X**2
         H=SQRT(X)*(TWO+Y*
    1 (P7(1)+Y*(P7(2)+Y*(P7(3)+Y*(P7(4)+Y*P7(5)))))/
    2 (Q7(1)+Y*(Q7(2)+Y*(Q7(3)+Y*(Q7(4)+Y*Q7(5))))))
        END IF
       ELSE IF(K .EQ. 1) THEN
        IF(X .LE. ONE) THEN
         Y=EXP(X)
         H=Y*(C2+Y*
    1 (P2(1)+Y*(P2(2)+Y*(P2(3)+Y*(P2(4)+Y*P2(5)))))/
    2 (Q2(1)+Y*(Q2(2)+Y*(Q2(3)+Y*(Q2(4)+Y*Q2(5))))))
        ELSE IF(X .LE. FOUR) THEN
         H=(P5(1)+X*(P5(2)+X*(P5(3)+X*(P5(4)+X*P5(5)))))/
    1 (Q5(1)+X*(Q5(2)+X*(Q5(3)+X*(Q5(4)+X*Q5(5)))))
        ELSE
         Y=ONE/X**2
         H=X*SQRT(X)*(C4+Y*
    1 (P8(1)+Y*(P8(2)+Y*(P8(3)+Y*(P8(4)+Y*P8(5)))))/
    2 (Q8(1)+Y*(Q8(2)+Y*(Q8(3)+Y*(Q8(4)+Y*Q8(5))))))
        END IF
       ELSE IF(K .EQ. 3) THEN
        IF(X .LE. ONE) THEN
         Y=EXP(X)
         H=Y*(C3+Y*
    1 (P3(1)+Y*(P3(2)+Y*(P3(3)+Y*(P3(4)+Y*P3(5)))))/
    2 (Q3(1)+Y*(Q3(2)+Y*(Q3(3)+Y*(Q3(4)+Y*Q3(5))))))
```

```fortran
      ELSE IF(X .LE. FOUR) THEN
        H=(P6(1)+X*(P6(2)+X*(P6(3)+X*(P6(4)+X*P6(5)))))/
     1  (Q6(1)+X*(Q6(2)+X*(Q6(3)+X*(Q6(4)+X*Q6(5)))))
        ELSE
          Y=ONE/X**2
          H=X**2*SQRT(X)*(C5+Y*
     1  (P9(1)+Y*(P9(2)+Y*(P9(3)+Y*(P9(4)+Y*P9(5)))))/
     2  (Q9(1)+Y*(Q9(2)+Y*(Q9(3)+Y*(Q9(4)+Y*Q9(5))))))
        END IF
      ELSE
        PRINT 100, NAME,K
        RETURN
      END IF
      IF(NAME .EQ. 'DFERDR') THEN
        GO TO 10
      ELSE
        FERDR=SNGL(H+(H-DBLE(SNGL(H))))
      END IF
      RETURN
      ENTRY DFERDR(DX,K)
      X=DX
      NAME='DFERDR'
      DFERDR=ZERO
      GO TO 9
   10 CONTINUE
      DFERDR=H
  100 FORMAT(7X,'***** CERN C323 ',A6,' ... INCORRECT K = ',I5)
      END
```

B.3 Coexistence.nb

The Mathematica notebook *Coexistence.nb* is a pedestrian but illustrative
way to determine the boundary of the coexistence region.

```
(* Program to determine the boundary of the coexistence
   region for the nuclear equation of state *)
(* Cell 1: - Evaluate once and proceed to Cell 2 *)
a2=21.1; a3=-38.3; a4=-26.7; a5=35.9; b01=0.692618;
b02 =0.0371196; b11 = -5.420171; b12 = 0.08182769;
b21 = 11.44679; b22 = -0.3117964; n0 = 0.15;
b0[T_] := b01 T + b02 T T;
b1[T_]:= b11 T + b12  T T;
b2[T_] := b21 T + b22  T T;
b[T_,x_]:=b0[T] + b1[T] x + b2[T] x^2;
p[T_,x_]:= (2/3) n0 a2(x/n0)^{5/3} +
  n0 a3 (x/n0)^{2}+(4/3) n0 a4 (x/n0)^{7/3}+(5/3) n0
    a5 (x/n0)^{8/3} + (2/3)  x b[T,x] ;
x[v_]:=1/v;
(* Cell 2: Evaluate isothermal pressure curves *)
   Plot[{p[0,x],p[8,x],p[14.54,x]},{x,0,.2}]
(* Cell 3:
   - Input T, a guess for the pressure pf, and evaluate.
   - Vary pf to have equal areas above and below curve.
   - Go to Cell 4 when a reasonable pf has been found *)
T = 13;
pf[x_]:=0.1845;
Plot[{pf[x[v]],p[T,x[v]]},{v,5,40}]
(* Cell 4: - Evaluate and check if integration = 0,
   - Vary pf in Cell 3 and evaluate again Cells 3 and 4,
   - Repeat until integration = 0 to a good approximation*)
v = 1/x/.NSolve[pf[x]-p[T,x]==0,x];
Integrate[pf[z]-p[T,x[z]],{z,v[[3]],v[[1]]}]
(* Cell 5:
   - Evaluate after integration=0 to a good approximation.
     These are the pressure, volumes and densities of the
     boundary at the given temperature *)
Print["Temperature=",T]
Print["Pressure=",pf[1]]
```

```
Print["Volumes=",v]
Print["Densities=",1/v]
Temperature=13
Pressure=0.1845
Volumes={33.9842,18.0896,10.3137}
Densities={0.0294255,0.0552803,0.0969584}
```

B.4   Percol.for

This fortran program performs bond percolation in 3-dimensions using random values of the bond probability. It reads an initial seed for the random number generator from the file percol.ran and updates the seed after every run of the program. The input data for the percolation is *nparl*, the linear size of the simple cubic lattice, and *nconf*, the number of percolation events to be generated. These data are stored on the file percol.inp.

The program percol.for will generate an output file named perr_xxx. dat with "xxx" equal to *nparl*, including information about the run and each of the percolation events calculated. It has the following structure:

50000 216	*50000 events*	*216 nodes*	
3 13 .49418	*3 sizes*	*13 fragments*	*probability=.49418*
9 3 1	*9 fragments*	*3 fragments*	*1 fragment*
1 2 201	*of size=1*	*of size=2*	*of size=201*
1 1 .98060	*1 size*	*1 fragment*	*probability=.98060*
1	*1 fragment*		
216	*of size=216*		
5 26 .36397	*5 sizes*	*26 fragments*	*probability=.36397*
21 1 2 1 1	*21 fragments*	*1 fragment*	*2 fragments ...*
1 2 3 5 182	*of size=1*	*of size=2*	*of size=5. . .*
⋮	⋮	⋮	⋮

As cryptically explained in the comments, percol.for outputs the number of total number of percolation events calculated, the number of bonds of the lattice and for each event, number of different fragment sizes, total number of fragments, bond breaking probability, and the number of fragments found for each different size and the fragment size.

```
program percol.for
implicit real (a-h,o-z)
double precision dseed
parameter(idi0=090,idi=(idi0+1)**3)
character*12 nombre
common/ilink/nparl,npar,iveci(6),q
open(90,file='percol.inp',status='old')
read(90,92)nparl,nconf
```

```
92    format(//,i6,i6,//)
      close(90)
      if(nparl.ge.idi0) then
        write(6,*) ' nparl must be < idi0 :',idi0
        stop
      end if
      open(95 ,file='percol.ran',status='old')
      read(95 ,*) dseed
      npar=nparl*nparl*nparl
      nparv=nparl+1
      iveci(1)=1
      iveci(2)=-1
      iveci(3)=nparv
      iveci(4)=-nparv
      iveci(5)=nparv**2
      iveci(6)=-nparv**2
      write(nombre,1113)nparl
1113  format('perr','\_',i3.3,'.dat')
      open(99,file=nombre,status='unknown')
9008  format(i6,i6,/)
      write(99,9008)nconf,npar
      do i0=1,nconf
        call clusters(dseed)
      end do
      close(99)
      rewind (95)
      write (95,*)dseed
      close (95)
      stop
      end

      subroutine clusters(dseed)
      implicit real (a-h,o-z)
      double precision dseed,random
      parameter(idi0=090,idi=(idi0+1)**3,id0=(idi0+1)**2)
      common/ilink/nparl,npar,iveci(6),q
      dimension ivaux(-idi:id0+idi),maclu(idi)
      dimension nclus(idi),nsalc(2,idi)
```

```
q=random(dseed)
numpc0=0
nfrag0=0
npar0=(npar1+1)
npar1=npar0**3
do i=(-npar1-1),npar1+npar1
  ivaux(i)=0
end do
do i=1,idi
  nclus(i)=0
  do ii=1,2
    nsalc(ii,i)=0
  end do
end do
iii=0
do i=1,nparl
  do j=1,nparl
    do k=1,nparl
      iii=iii+1
      ivaux(iii)=iii
    end do
    iii=iii+1
    ivaux(iii)=0
  end do
  do l=1,nparl+1
    iii=iii+1
    ivaux(iii)=0
  end do
enddo
numparc=0
ic=0
do 1001 i1=1,npar1
  if(ivaux(i1).ne.0) then
    ic=ic+1
    do i=i,npar
      maclu(i)=0
    end do
    numpc=1
```

```
      numpa=0
      maclu(numpc)=ivaux(i1)
      ivaux(i1)=0
   else
      go to 1001
   end if
   numparc=numparc+1
   do while(numpa.lt.numpc)
      numpa=numpa+1
      indana=maclu(numpa)
      do i2=1,6
         indipar=indana+iveci(i2)
         ipar=ivaux(indipar)
         ilink=0
         if(ipar.ne.0) then
            if(q.gt.random(dseed)) then
               ilink=1
               numpc=numpc+1
               numparc=numparc+1
               maclu(numpc)=ipar
               ivaux(indipar)=0
            end if
         end if
      enddo
   enddo
   numpc0=numpc0+numpc
   nfrag0=nfrag0+1
   nclus(numpc)=nclus(numpc)+1
1001 continue
   numpc0=0
   numpc00=0
   indicl=0
   do inum=1,idi
      if(nclus(inum).ne.0) then
         indicl=indicl+1
         numpc0=numpc0+nclus(inum)*inum
         nsalc(1,indicl)=nclus(inum)
         nsalc(2,indicl)=inum
```

```
      end if
    end do
    write(99,9009)indic1,ic,q
    write(99,9010)(nsalc(1,jj),jj=1,indic1)
    write(99,9010)(nsalc(2,jj),jj=1,indic1)
9009  format(2(i8),f14.8)
9010  format(40(i6))
    return
    end

    Double precision Function random(dseed)
    double precision dseed
    double precision d2p31m,d2p31
    data d2p31m/2147483647.d0/
    data d2p31/2147483711.d0/
    dseed=dmod(16807.d0*dseed,d2p31m)
    random=dseed/d2p31
    return
    end
```

B.5   Percol.inp

This file contains the input data for percol.for with the format nparl,nconf
each integer numbers in 6-digit fields. nparl is the linear size of the simple
cubic lattice to be used, and nconf is the number of percolation events to
be generated.
```
nparl,nconf
123456 23456
    06 10000
```

B.6   Percol.ran

This file contains the initial seed for the random number generator used in
percol.for. It is updated after every run of the program.
```
    1.00712374E+09
```

Bibliography

Aichelin J. (1991) *Phys. Rep.* **202**, 233.

Albergo S. *et al.* (1985) *Il Nuovo Cimento* **A89**, 1.

Aranda A. and López J. A. (1995) *Rev. Mex. Phys.* **41**, 90.

Ayik S. *et al.* (1996) *Z. Phys.* **A355**, 407.

Barber M. N. (1978) in *Phase Transitions and Critical Phenomena* Eds. Domb C. and Green M.S., Vol. 8, 145-266.

Barrañón A., Chernomoretz A., Dorso C. O. and López J. A. (1999) *Rev. Mex. Fis.* **45(S2)**, 110.

Bauer W. (1998) *Phys. Rev.* **C38**, 1297.

Baumgardt H. G. *et al.* (1975) *Z. Phys.* **A273**, 359.

Beck C. and Schlogl F. (1997) *Thermodynamics of chaotic systems*, Cambridge University Press, Cambridge.

Belkacem M., Latora V. and Bonasera A. (1995) *Phys. Rev.* **C52**, 271.

Benettin G., Galgani L. and Strelcyn J. M. (1976) *Phys. Rev.* **A14**, 2338; Benettin G., Galgani L., Giorgilli A. and Strelcyn J. M. (1980) *Meccanica* Mar., 21.

Bertsch G. F. and das Gupta S. (1988) *Phys. Rep.* **160**, 189.

Bertsch G. F. and Siemens P. (1984) *Nuc. Phys.* **A314**, 465.

Binney J. J., Dowrick N. J., Fisher A. J. and Newman M. E. J. (1995) *The Theory of Critical Phenomena, an introduction to renormalization group*, Oxford Science Publications, Oxford.

Boal D. H. and Glosli J. N. (1988) *Phys. Rev.* **C38**, 1870; Boal D. H., Glosli J. N. and Wicentowich C. (1989) *Phys. Rev. Lett.* **62**, 737.

Bohr A. and Mottelson B. (1973) *Ann. Rev. Nuc. Sci.* **23**, 363.

Bondorf J. P. *et al.* (1995) *Phys. Rep.* **257**, 133; Bondorf J. P. *et al.* (1985) *Nuc. Phys.* **A444**, 460.

Borderie B. *et al.* (1990) *Eur. Phys. J.* **A6**, 197.

Botet R. and Ploszajczak M. (1994) *Int. J. Mod. Phys.* **E4**, 1033.

Botvina A. S., Iljinov A. S., and Mishustin I. N. (1990) *Nuc. Phys.* **A507**, 649.

Bravina L. V. and Zabrodin E. E. (1996) *Phys. Rev.* **C54**, R464.

Cahn J. W. (1961) *Acta Metall* **9**, 795.

Campi X. (1986) *J. Phys.* **A19**, L91;
    Campi X. (1986) *J. Phys.* **C2** 48 C2-151;
    Campi X. (1986) *Phys. Lett.* **B208** 351.

Campi X. and Krivine H. (1986) *Z. Phys.* **344** 81.

Campi X. and Krivine H. (1994) in *Dynamical Features of Nuclei and Finite Fermi Systems*, World Scientific, Singapore.

Campi X. and Desbois J. (1985) in *Proceedings of the 23rd Bormio Conference*, Ric. Sci. Educ. Perm., Milano, 497.

Charity R. J. (1988) *Nuc. Phys.* **A483** 371.

Chernomoretz A., Dorso C. O. and López J. A. (2000) *in print, Heavy Ion Phys.* **11**, 351.

Cody W. J. and Thacher H. S. (1967) *Math. Comp.* **21**, 30.

D'agostino M. *et al.* (1995) *Phys. Rev. Lett.* **75**, 4373.

Danielewicz P. (1973) *Nuclear Physics* **A314**, 465.

Desbois J. (1987) *Nucl. Phys.* **A466**, 724.

Domb C. and Green M. S. Eds. (1980) *Phase Transitions & Critical Phenomena*, Vol. *2*, Pergamon Press Ltd., New York.

Dorso C. O. and Donangelo R. (1990) *Phys. Lett.* **B244**, 165;
    Dorso C. O., Balonga P. and Donangelo R. (1993) *Phys. Rev.* **C47**, 2204;
    Donangelo R., Dorso C. O. and Marta H. (1991) *Phys. Lett.* **B263**, 14.

Dorso C. O. and Randrup J. (1993) *Phys. Lett.* **B301**, 328;
    Dorso C. O. and Aichelin J. (1995) *Phys. Lett.* **B345**, 197;
    Reposeur T., Sebille F., de la Mota V., and Dorso C. O., (1997) *Z. Phys.* **A357**, 79.

Dorso C. O. and Randrup J. (1987) *Phys. Lett.* **B188**, 287;
    Dorso C. O. and Randrup J. (1988) *Phys. Lett.* **B215**, 611;
    Dorso C. O. and Randrup J. (1989) *Phys. Lett.* **B232**, 29.

Dorso C. O., Latora V. and Bonasera A. (1999) *Phys. Rev.* **C60**, 1.

Selected Durand (1998) Aspects of the Physics of Hot Nuclei, Preprint LPC Caen **LPCC 98-02**.

Hauger J. A. *et al.* (1996) *Phys. Rev. Lett.* **77**, 235.

Elliot J.B. *et al.* (1997) *Phys. Rev.* **C55**, 1319.

Fabris D. *et el.* (1987) *Phys. Lett.* **B196**, 429.

Finn J. E. *et al.* (1982) *Phys. Rev. Lett.* **49**, 1321.

Fisher M. E. (1971) in *Critical Phenomena, Proceedings of the International School of Physics "Enrico Fermi" Course 51*, Ed. Green M. S., Academic, New York, 1.

Ford J. in Cohen E. D. G. Ed. (1975) *Fundamental Problems in Statistical Mechanics, Vol. 3*, North Holland, Amsterdam.

Friedman W. A. (1990) *Phys. Rev.* **C42**, 667.

Friedman W. A. and Lynch W.G. (1983) *Phys. Rev.* **C28**, 28.

Gilkes M. L. *et al.* (1994) *Phys. Rev. Lett.* **73**, 1590.

Goldenfeld N. (1992) *Lectures on Phase Transitions and the Renormalization Group* Addison-Wesley, Massachusetts.

Gómez del Campo J. *et al.* (1979) *Phys. Rev.* **C19**, 2170.

Goodman A. L., Kapusta J. I. and Mekjian A. Z. (1984) *Phys. Rev.* **C30**, 851.

Gross D. H. E. (1997) *Phys. Rep.* **279**, 119;
   Gross D. H. E. (1984) *Rep.Prog. Phys.* **53**, 605.

Harmon B. A., Pouliot J., López J.A., Suro J., Knop R., Chan Y., DiGregorio D. E., and Stokstad R. G. (1990) *Phys. Lett.* **B235**, 234.

Hillert M. (1961) *Acta Metall* **9**, 525.

Hirsch A. S. *et al.* (1984) *Phys. Rev.* **C29**, 508.

Huang K. (1980) *Statistical Mechanics*, John Wiley, New York.

Jakobsson B. *et al.* (1977) *Nucl. Phys.* **A276**, 523.

Kadanoff (1966) *Physics* **2**, 263.

Kapusta J. (1984) *Phys. Rev.* **C29**, 1735.

Koonin S. E. and Randrup J. (1987) *Nucl. Phys.* **A474**, 173.

Kwiatkowski K. *et el.* (1998) *Phys. Lett.* **B423**, 21.

Labastie P. and Whetten R. L. (1990) *Phys. Rev. Lett.* **65**, 1567.

Landau L. D. and Lifshitz E. M. (1980) *Statistical Physics, 3rd Ed., Part 1*, Pergamon Press Ltd, New York.

Li T. *et al.* (1994) *Phys. Rev.* **C49**, 1630.

Lichtemberg A. J. and Lieberman M.A. (1992) *Regular and chaotic Dynamics*, Springer Verlag.

López J. A. and Lübeck G. (1989) *Phys. Lett.* **B219**, 215.

López J. A. and Randrup J. (1989) *Nuc. Phys.* **A503**, 183.

López J. A. and Randrup J. (1990) *Nuc. Phys.* **A512**, 345.

López J. A. and Randrup J. (1994) *Nuc. Phys.* **A571**, 379;
   López J. A. (1996) *Heavy Ion Phys.* **3**, 141.

López J. A. and Randrup J. (1989) *Nuc. Phys.* **A491**, 477;
   López J. A. (1992) *Rev. Mex. Fis.* **38**, 95.

López J. A. and Randrup J. (1992) *Comp. Phys. Comm.* **70**, 92.

López J. and Siemens P. (1984) *Nuc. Phys.* **A314**, 465.

Mastinu P. F. *et al.* (1996) *Phys. Rev. Lett.* **76**, 2646.

McQuarrie D. (1973) *Statistical Mechanics*, Harper and Row, New York.

Nayak S. K., Ramaswamy R., and Chakravarty C. (1995) *Phys. Rev.* **E51**, 3376;
   Mehra V. and Ramaswamy R. (1997) *Phys. Rev.* **E56**, 2508;
   Tanner G. M., Bhattacharya A., Nayak S. K. and Mahanti S. D. (1997) *Phys. Rev.* **E55**, 322.

Ogilvie C. A. *et al.* (1991) *Phys. Rev. Lett.* **67**, 1214.

Oseledec V.I. (1968) *Trans. Moscow Math. Soc.* **19**, 2336.

Panagiotou A. D. *et al.* (1984) *Phys. Rev. Lett.* **52**, 496.

Lenk R. J., Schlagel T. J. and Pandharipande V. R. (1990) *Phys. Rev.* **C42**, 372.;
   Lenk R. and Pandharipande V. R. (1986) *Phys. Rev.* **C34**, 177.;
   Schlagel T. J. and Pandharipande V. R. (1987) *Phys. Rev.* **C36**, 162.

Peilert G., Konpka J., Stocker H., Greiner W., Blann M, and Mustafa M. G.

(1992) *Phys. Rev.* **C46**, 1457.

Pethick C. J. and Ravenhall D. G. (1987) *Nucl. Phys.* **A471**, 19;
Pethick C. J. and Ravenhall D. G. (1986) in *Hadrons in Collision* Ed.
Carruthers P. and Strottman D., World Scientific, Singapore, 277;
H. Heiselberg, Pethick C. J. and Ravenhall D. G. (1988) *Phys. Rev. Lett.*
61, 818.

Phair L., Bauer W. and Gelbke C. K. (1993) *Phys. Lett.* **B314**, 271.

Pochodzalla J. *et al.* (1995) *Phys. Rev. Lett.* **75**, 1040.

Porile N. T. *et al.* (1989) *Phys. Rev.* **C39**, 1914.

Pouliot J., Beaulieu L., Djerroud B., Dore D., Laforest R., Roy R., St-Pierre C.
and López J. A. (1993) *Phys. Rev.* **C48**, 2514.

Pratt S., Montoya C. and Ronning F. (1995) *Phys. Lett.* **B349** 261.

Randrup J. and Ayik S. (1994) in *International Workshop on the Dynamical
Features of Nuclei and Finite Fermi Systems.*, World Scientific, Singapore,
64.

Rivet *et al.* (1999) in *Proceedings of the XXVII International Workshop on Gross
Properties of Nuclei and Nuclear Excitations.*, Hirschegg, Austria, January
1999.

Samaddar S. K., De J. N. and Shlomo S. (1997) *Phys. Rev. Lett.* **79**, 4962.

Segrè E. (1982) *Nuclei and Particles*, 2nd Ed., Addison Wesley, Massachusetts.

Serfling V. *et al.* (1998) *Phys. Rev. Lett.* **80**, 3928.

Shimada I. and Nagashima T. (1979) *Prog. Theor. Phys.* **61**, 1605.

Stanley H. E.(1971) *Introduction to Phase Transitions and Critical Phenomena*,
Oxford University Press, Oxford.

Stauffer D. and Aharony A. (1992) *Introduction to Percolation Theory*, Taylor
and Francis, London.

Strachan A. and Dorso C. O. (1997) *Phys. Rev.* **C55**, 99.

Strachan A. and Dorso C. O.(1997) *Phys. Rev.* **C55**, 775.

Strachan A. and Dorso C. O. (1999) *Phys. Rev.* **C59**, 285.;
(1998) *Phys. Rev.* **C58**, R632.

Trockel R. *et al.* (1989) *Phys. Rev.* **C39**, 729.

Tsang M. B. *et al.* (1993) *Phys. Rev. Lett.* **71**, 1502.

Tsang M. B., Lynch W. G., Xi H. and Friedman W. A. (1997) *Phys. Rev. Lett.*
78, 3836.

Walecka J. D. (1995) *Theoretical Nuclear and Subnuclear Physics*, Oxford University Press, New York.

Wang G. *et al.* (1999) *Phys. Rev.* **C60**, 14603.

Warwick *et al.* (1983) *Phys. Rev.* **C27**, 1083.

Weisskopf V. F. (1937) *Phys. Rev.* **52**, 295.

Widom B. (1965) *J. Chem. Phys.* **43**, 3892.

Wilets L., Yariv Y. and Chestnut R. (1978) *Nuc. Phy.* **A301**, 359.

Yeomans J. M. (1994) *Statistical Physics of Phase Transitions*, Clarendon Press,
Oxford.

Youngblood D. H., Rozsa C. M., Moss J. M., Brown D. R., and Bronson J. D.
(1977) *Phys. Rev.* **39**, 1188.

Index